FRANCO VALENTE
BRITISH GAS EXPLORATION & PRODUCTION
FED READING

VALVE USERS MANUAL

A technical reference book on industrial valves for the control of fluids

Edited by J. Kemplay CEng,FIMechE

MECHANICAL ENGINEERING PUBLICATIONS LTD
LONDON

© BRITISH VALVE MANUFACTURERS' ASSOCIATION 1980

ISBN 0 85298 428 6

Phototypeset by Santype International Limited, Salisbury, Wiltshire.
Printed and bound in Great Britain by
The Burlington Press (Cambridge) Ltd.,
Foxton, Royston, Hertfordshire.

CONTENTS

British Valve Manufacturers' Association

Established in 1939, the British Valve Manufacturers' Association is the national representative body of the British manufacturers of industrial valves for the control of fluids. The Association represents the British Valve Industry at the highest level in such matters as Government legislation and liaison with Government economic and technical departments; the determination of international and British Standards; relations with other industrial bodies both internationally and nationally; industrial publicity, and other subjects affecting commercial policy.

Membership of the Association is open to British companies which, wholly or to a considerable extent, manufacture industrial valves for the control of fluids and actuators therefor. Membership by such a company signifies that company's willingness to work for the common good of the valve industry and its customers, and may be taken as an indication of a progressive, forward looking enterprise. The Association, governed by its Executive Committee, has permanent committees for technical and marketing requirements: the Technical Committee which considers matters relating to the design, production, installation, testing and maintenance of valves and which handles standardization and technical liaison; and the Marketing Committee which promotes the products of member firms by the distribution of technical reference books, films, learned papers, and participation in Exhibitions, and undertakes market analysis and institutes market research. In addition to sponsoring this publication the BVMA also has available a buyers guide entitled 'Valves from Britain'.

3 Pannells Court, Chertsey Street,
Guildford, Surrey GU1 4EU

Tel: Guildford (0483) 37379

Preface

This volume has been preceded by three editions of a 'Technical Reference Book on Valves for the Control of Fluids'. The present publication is a completely new work which it is hoped will be of even greater value to the valve user—hence the title.

Our aim has been to provide as much information as possible on the various design aspects and features of the most commonly used valves and their variants. At the same time background data is provided on most other aspects which impinge on valve selection and use.

We are conscious that at the date of publication the United Kingdom has still not fully completed its move to the use of the metric system of units but as we in the Association and the industry as a whole are fully committed to metrication we have written in these terms but have provided conversion tables.

Because of the interim stage in metrication in the U.K. as a whole and the continual developments in national and international Standards (in which the Association plays a major role) we have generally avoided reference to any specific Standards and have considered it wiser to indicate sources for data to enable the reader to ensure that he is up-to-date, at any time.

The preparation of a book of this kind is a major undertaking and has involved considerable time and effort by a large number of our members both in committees and in preparing data individually. I have great pleasure in thanking them all for their contributions as well as our friends in supporting industries who have assisted in making the book complete. We are also indebted to the Science Museum, London for allowing us to use certain illustrations and data.

Chairman
British Valve Manufacturers' Association Ltd.

HISTORY — a brief survey

It is a cruel mortification in searching for what is instructive in the history of the past times, to find that the exploits of conquerors who have desolated the earth and the freaks of tyrants who have rendered nations unhappy, are recorded with minute and often disgusting accuracy – while the discovery of the useful arts and the progress of the most beneficial branches of commerce are passed over in silence and suffered to sink into oblivion. (From William Robertson's historical disquisition on Ancient India.)

The quotation from Robertson's *India* expresses very effectively the situation in respect to the early development of many commonplace items in the field of engineering. Pipe, valves and their associated fittings are no exception and the paucity of our knowledge of man's early efforts in this field is, as Robertson so aptly puts it, cruelly mortifying.

History does give us a few gleanings. Pipe in the earliest conception was made of various materials such as earthenware, stone, wood,[1] leather, and, later on, lead and copper. Ancient records mention pipe of sheet lead in the old cities of Asia, Egypt, and Greece. The Romans made great use of lead pipes, which were formed in short lengths by shaping cast lead sheet around a wooden former and soldering the seam. Cast lead pipes were first produced during the sixteenth century in England and it was not until late in the eighteenth century that Wilkinson, the English iron master, took out a patent for drawing lead pipes through dies which led to the development of the same process for producing pipes from other materials.

Prior to Greek and Roman times little is known about the methods used to control the flow of fluids. Doubtless some form of sluice gate was used to hold and retain water in irrigation channels and we know there was some knowledge of the principles of flow because of the water clocks made by the early Egyptians. Probably the most sophisticated device of the period was the bellows, since this involved the use of leather flap valves, operated either manually or automatically.

The Greek and Roman periods, roughly 600 BC to AD 400, saw the development of many mechanical and hydraulic machines and the first use of valves of sophisticated design. In one case, the plug cock valve, the basic design remained virtually unchanged until the nineteenth century.

Flap valves and coin valves, the forerunners of the present day swing and lift check valves, were being used in water force pumps, and bronze or brass plug cocks were in common use as stop valves on water mains and supply pipes to public and domestic buildings. Several fine examples of these valves have survived and can be seen in the museum at Naples and the British Museum. The large bronze cock valve shown in Fig. 1, preserved at Naples, was found at Capri among the ruins of the palace of Tiberius built about AD 25.

We know very little about the development of valves during the thousand years following the Roman period. From the early sixteenth century comes some evidence in the form of a few sketches in one of Leonardo da Vinci's notebooks. These show several designs of guided cone disk lift valves (Fig. 2) and multiple clapper valves but we do not know whether these valves were ever actually made.

An interesting illustration, Fig. 3, from Giovanni Bronca's *Machines*, Rome, 1629, shows the 'grinding in' of a taper plug cock by machinery and in the explanation is a note that 'in the latter part of the fifteenth century it became essential to construct valves (cocks) in a more careful manner for use on pneumatic pumps'. In 1681 Dennis Papin invented the lever type of safety valve for use on his 'digester' or pressure vessel and in 1718 Jean Disaquliers adapted Papin's valve for use as a safety valve on steam boilers.

The early years of the eighteenth century marked the arrival of the steam engine as a practical and commercial proposition. In 1698 Thomas Savery had patented his engine for the 'raising of water' and in 1705 Thomas Newcomen introduced his advanced version of Savery's engine, the atmospheric beam engine.[2] James Watt provided the decisive step forward in the development of the steam engine when in 1769 he patented the separate condenser.

From the illustrations and descriptions of these engines and other steam engines of the period it appears that several types of valve were used. These included plug cock valves, manually operated lift valves, automatic lift valves (lift check), and flap valves (swing check) (Fig. 4). Watt introduced a throttle valve to his engines. This was basically what is now called a butterfly valve and consisted of a pivoted circular plate or disk fixed into a pipe and linked to the engine governor (Fig. 5).

The plug cock predominated as a stop valve until the end of the eighteenth century, when Maudsley's improved screw-cutting lathe facilitated the production of the screw-down stop valve. The design of screw-down stop valve or so-called globe valve, that we know today, appears to have been introduced in the period 1840–50.

During the nineteenth century a number of eminent engineers directed their attention to valves. One of these was Timothy Hackworth, an associate of George Stephenson and builder of the famous locomotive 'Royal George'. He introduced adjustable springs instead of weights to the steam safety valve. Figure 6 shows the spring arrangement of one of his valves, now preserved in the Science Museum, London.

In 1839 James Nasmyth, whose greatest achievement was probably the invention of the steam hammer, patented a double faced wedge shaped disk gate valve for main street water pipes. This valve incorporated features that still form the basis of wedge gate valve practice (Fig. 7).

Another major innovation was the introduction of the groove-packed plug cock by Dewrance & Co. in 1875. Grooves around the body portways were filled with asbestos, forming slightly raised rings. This made the

Fig. 1. Ancient bronze cock

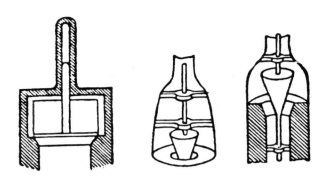

Fig. 2. Valves from Leonardo da Vinci, 1514

Fig. 3. Grinding of
cocks, 1629

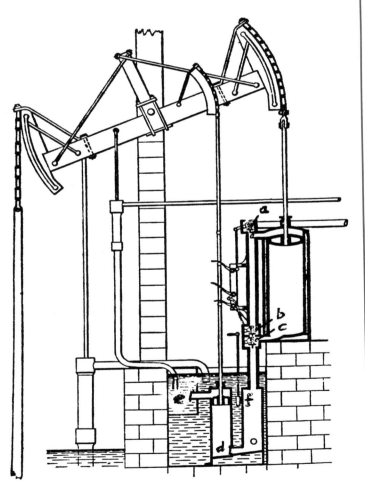

Fig. 4. Watt's single-acting engine 1769
a, b, c – Lift valves
d, e – Flap valves
f – Plug cock valve

valve easier to operate and more suitable for use with steam.

Some eleven years later, in 1886, Joseph Hopkinson introduced the parallel slide gate valve in which a two-piece parallel faced flexible disk was the key feature and sealing was effected by the action of the line pressure on the disk. The parallel slide valve of the present day is still based on Hopkinson's idea.

The great technological advances in all spheres of industrial activity during the present century have presented the valve industry with a host of new requirements. Today almost every conceivable fluid is handled in pipe and operating conditions have been extended at both ends of the scale. Pressures now range from near vacuum to upwards of 200 bar and thermal conditions from cryogenic to temperatures in excess of 600°C. Additionally, there is now an increasing number of special operational problems to be overcome, e.g., certain applications in the field of nuclear power, and these often necessitate the design and manufacture of purpose made valves.

These challenges to the valve industry have been met and overcome in various ways. Traditional valve types such as gate and globe valves have been reappraised in the light of the latest advances in metallurgy and the availability of new plastics and synthetic rubbers and with full recognition of the design and quality improvements offered by modern manufacturing techniques. Standard designs have been improved and refined to obtain the highest possible levels of performance and many innovations have been introduced to provide additional features and qualities.

In addition to the traditional valves many other types of valves, notably lubricated taper plug, diaphragm, ball, and butterfly valves, have been developed and engineered into practical and industrially acceptable products. Little known thirty or forty years ago, these valves have since found increasing favour for a great variety of industrial applications and today they hold substantial shares of the market for general purpose valves. It seems fitting to conclude this dissertation with a few notes on the historical development of these valves.

Lubricated Taper Plug Valve

This was introduced by a Swedish engineer, Sven Nordstrom, during the time of World War I. Searching for some means of overcoming the excessive leakage and sticking which he had encountered with ordinary taper plug valves, he devised a method of forcing lubricant between the sealing surfaces of the body and the plug. This overcame the sticking problem and, because the lubricant was under a greater pressure than the service fluid, it also provided a seal and greatly reduced leakage.

During the sixty years or so since its inception Nordstrom's original design has been subjected to considerable development, one outcome being the introduction of a high pressure version of the valve in the late nineteen-thirties. More recently, the employment of newly developed multi-purpose lubricants has extended both the temperature ceiling and the field of application of the valve.

Diaphragm Valve

Although the principle of the diaphragm valve has been known for a long time it was not developed as an individual type of valve with specific design features and characteristics until the early part of the present century. At that time engineers in the gold mines of South Africa were faced with the problem of excessive leakage of com-

pressed air at the glands of the valves then being used. A solution was found by a South African engineer named Saunders, who in 1929 devised a valve in which a flexible rubber diaphragm was used as the closing member. Moreover, the diaphragm was arranged in such a manner that it isolated the valve operating mechanism from the compressed air, thereby eliminating any possibility of leakage in that direction.

The valve of 1929 soon became a great success and since that time it has been developed and put to many other uses. Once a special valve designed for the sole purpose of handling compressed air, the diaphragm valve is now fully recognized as a valve type with proven application in a variety of services in many fields of industry.

Ball Valve

The ball valve or spherical plug valve is a relative newcomer to the valve family. Although known and made during the period 1920/1939 it was given little attention until the early nineteen-forties when it was found to be well suited, as a valve type, for use in the fuel systems of new aircraft then being designed. The existing designs were subjected to intensive development to meet the exacting requirements of the aircraft industry and an original design resulted that incorporated a 'floating' ball and included synthetic rubber seals. This work paved the way for further development in the immediate postwar years which eventually produced the first industrial ranges of ball valves.

During the last twenty years many valve manufacturers have directed their attentions to the ball valve and a variety of new improved designs have been introduced, a major factor in the success of these developments being the use of new types of plastics and elastomers for the seatings and seals. This has led to a much wider diversification and expansion of the capabilities of the ball valve for duties in practically all sections of the valve market.

Butterfly Valve

Before the beginning of the present century the butterfly valve, in elementary form, had been used for mostly simple duties such as dampers in ducting and engine throttle valves. Reference to the use by James Watt of such a device has already been made and much later the ancestor of the modern motor car, the first Mercédès built around the year 1901, introduced a carburetter which had a butterfly valve linked to the accelerator pedal.

It was probably around this latter time that butterfly valves constructed along the lines we know today first came into use for pipeline duties. At first only metal to metal seats were available but it was not long before rubber seats were introduced to improve the fluid tightness of the valve and it then began to find greater employment.

Since World War II, design improvements and the use of modern synthetic rubbers for the sealing members have extended the application of the butterfly valve into many industrial fields for which it was not previously considered.

---

[1] It is interesting to note that an article in the *Chartered Mechanical Engineer*, June 1977 issue, gives consideration to wood as an alternative material for future gas and oil pipelines.

[2] Beam engines were made for over two hundred years and it is interesting that the last beam type of steam pumping engine to be produced in the United Kingdom was actually built in 1919 by a valve manufacturer, Glenfield & Kennedy Ltd. (now Neptune Glenfield Ltd).

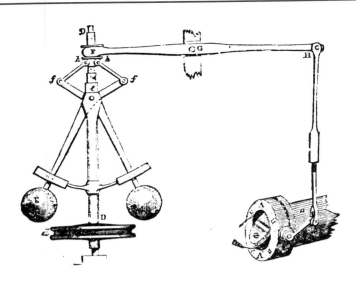

Fig. 5. Watt's governor and throttle-valve

Fig. 6. Spring arrangement of Hackworth's safety valve (Photo, Science Museum, London)

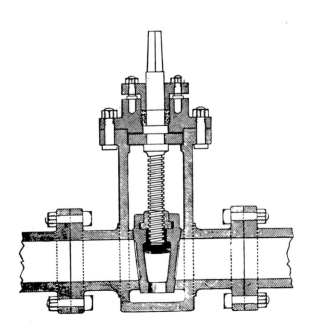

Fig. 7. Nasmyth's double faced, wedge-shaped, sluice-valve

TYPES OF VALVES

BALL VALVE

Some confusion still exists concerning the proper terminology for this valve. The circumstances causing this are (1) that the valve is basically a member of the plug valve family, and (2) that the term 'ball valve' is frequently used when referring in fact to the ball float operated valve. Names such as 'spherical plug valve' and 'ball plug valve' are still sometimes used but the simple term 'ball valve' has now become dominant and will be used here.

The ball valve is an adaptation of the plug valve. It has the same 90° rotary movement and the rotary member is in constant contact with the seats. The plug is in the form of a ball with a circular hole or flow way through one axis. The proportions of the ball and flow way are such that when the ball is given a quarter turn a full spherical face is presented to the inlet and outlet ports of the valve, thus shutting off the flow.

Ball valves offer positive tight shut-off with a quarter turn operation and low operating torque. Full parallel bore valves provide straight through flow with a minimum of resistance and even with reduced bore valves the pressure drop across the valve is still extremely small. Consequently, and for economic reasons, the majority of ball valves today have a reduced bore through the ball, the most common exception to this being when some

form of 'pigging' of the pipeline is required.

Ball valves are generally considered to be most suited for straight on–off duties but recent developments have introduced ball valves specially designed for throttling and flow control.

An important aspect of the ball valve is the inherent compactness of the design, making for easy handling and maintenance.

Most standard ball valves have an operating temperature range of between $-30\,°C$ and $230\,°C$ at pressures from a coarse vacuum (25 torr) to 51 bar, depending on size. Specialized valves are available, however, for services below $-200\,°C$ and above $500\,°C$ and from very high vacuum (10^{-9} torr) to above 400 bar.

The applications of ball valves are as wide and varied as industry itself. They range from simple services such as water, solvents, acids, and natural gas to more difficult and dangerous services such as gaseous oxygen, hydrogen peroxide, methane, and ethylene. Limitations of use are governed by the temperature and pressure characteristics of the seat material, but research and development of new materials make it foreseeable that the future will see an extended use of the ball valve for an even wider range of applications than at present.

BODY TYPES

These may be divided into one-piece bodies and multi-piece bodies and Fig. 1 illustrates the variations available.

The one-piece body type provides a very rigid construction. In the top entry version (Fig. 2), it is possible to remove the ball and seats from the valve without taking the valve out of the line, an advantage where maintenance in-situ is permitted and where valves are welded into the line. The one-piece body with end entry (Fig. 3) provides a compact design which dispenses with the need for a body joint, hence eliminating a potential leak path. Here the valve must be removed from the

pipe line to obtain access to the working parts. This can be an essential requirement in some industries where it is often forbidden to do maintenance in-situ.

Multi-piece bodies, two-piece or three-piece (sandwich), offer greater scope for ingenuity of design. For example, in the sandwich type (Fig. 4), the central portion containing all the working parts can be removed as a unit from the valve, leaving the two body end connectors in position in the line. Again, this is useful in the case of welded ends and it also makes possible the provision of interchangeable end connectors.

SEATING MATERIALS

Although ball valves with all metal seats are used on certain applications, the most general seating combination is a metal ball contained between plastic or elastomer seatings. Polytetrafluorethylene (PTFE), either virgin or

'filled', is commonly used for seats but a number of other materials are used from time to time for particular applications. These include nylon, 'Buna N', graphite and PCTFE.

BALL SUPPORT DESIGN AND SEALING

Two design concepts are used for supporting the ball, namely floating ball and trunnion mounted. In the floating ball arrangement the ball is supported by the seatings but is free to float laterally when the valve is closed. Sealing is obtained by the action of the line pressure urging the ball against the downstream seating. In the trunnion mounted design the ball is held in a fixed position top and bottom by trunnions and bearings. The seal is made by the line pressure acting on the back of the

upstream seat ring and pressing it against the ball.

For applications where the line pressure is insufficient to ensure a positive leak-tight seal some kind of built-in seat loading is incorporated. This may take the form of springs arranged behind the seatings or the seatings themselves may be preloaded on assembly. In another method the seatings are specially shaped so as to obtain a designed deformation which ensures a seat loading force irrespective of line pressure.

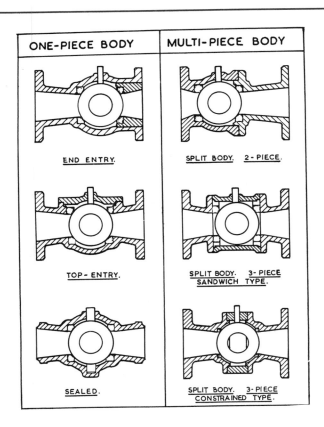

ONE-PIECE BODY	MULTI-PIECE BODY
END ENTRY.	SPLIT BODY. 2-PIECE.
TOP-ENTRY.	SPLIT BODY. 3-PIECE SANDWICH TYPE.
SEALED.	SPLIT BODY. 3-PIECE CONSTRAINED TYPE.

Fig. 1. Ball valve constructional types

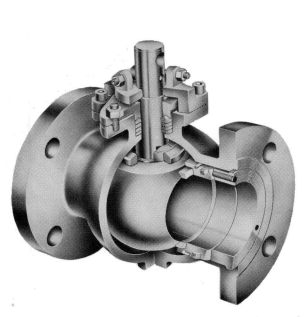

Fig. 3. Ball valve with one-piece body, end entry and floating ball. Full bore illustrated, reduced bore also available

Fig. 2. Ball valve with one-piece body, top entry and trunnion mounted ball

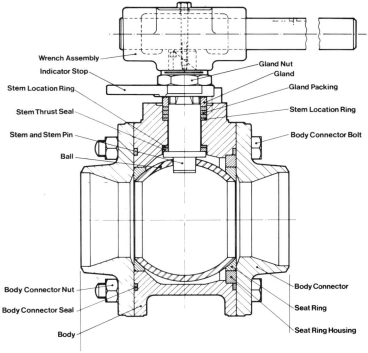

Wrench Assembly
Indicator Stop
Stem Location Ring
Stem Thrust Seal
Stem and Stem Pin
Ball
Gland Nut
Gland
Gland Packing
Stem Location Ring
Body Connector Bolt
Body Connector Nut
Body Connector Seal
Body
Body Connector
Seat Ring
Seat Ring Housing

Fig. 4. Ball valve with three-piece (sandwich) body and floating ball

13

BUTTERFLY VALVE

The name comes from the wing-like structure and action of the valve disk which is arranged on the pipe diameter principle. The circular shaped disk turns about a diametrical axis within the cylindrical bore of the valve body and a quarter turn rotation of the disk opens or closes the valve.

The basic simplicity of the design provides a compact and relatively low weight valve having few component parts. The quarter turn action offers quick opening or closing with ease of operation and the valve has good flow control characteristics.

In the wide open position the only obstruction to flow is that due to the thickness of the disk, so the pressure drop across the valve is small. During the closing movement the rate of cut-off of the flow diminishes as the disk moves towards the closed position, making the valve well suited for flow regulation purposes.

Butterfly valves are usually either resilient or metal to metal seated. Resilient seated valves may have the resilient seating arranged in the body or attached to the periphery of the disk.

Metal seated valves will normally provide longer life than resilient seated designs but are more difficult to make completely leak-tight. They are better suited for higher temperature duties as they do not have the temperature restrictions imposed by resilient seatings.

An extensive range of synthetic elastomer and plastic materials for use as resilient seatings has been developed as a result of close co-operation between valve designers and specialists in the metallurgical, rubber, and plastics industries. Positive shut-off with repeatability of performance is assured and the wide choice of seating materials provides the butterfly valve with an expansive range of service applications.

If the valve is required for flow control purposes it is essential that there should be correct selection of size and type to give satisfactory and effective operation.

The design concept of the butterfly valve is particularly well suited to large size manufacture. Sizes up to 10 metres diameter have been produced and even larger sizes are presently being considered for tidal and ocean thermal energy conversion schemes. In general industrial applications butterfly valves have found wide acceptance in the oil, gas, chemical, water treatment and process industries and are used in the condenser and circulating water systems of thermal power stations.

TYPES

Wafer
A valve for clamping between pipe flanges using through bolting. The body may be 'single flange', flangeless, or 'U' section, as illustrated in Figs 2, 3, and 4, respectively.

Double Flanged
A valve having flanged ends for connection to pipe flanges by individual bolting (Fig. 1).

A wafer type of valve with a 'U' section body will also come within this category if it is suitable for the individual bolting of each flange to the pipework.

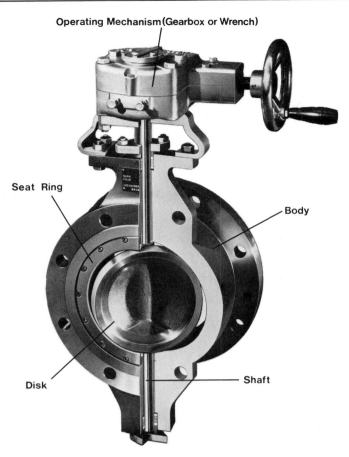

Operating Mechanism (Gearbox or Wrench)

Seat Ring

Body

Disk

Shaft

Fig. 1. Butterfly valve, double flanged, fitted with gear operating unit

Fig. 2. Butterfly valve, wafer/single flange, with rubber lined body. Fitted with gear operating unit

Fig. 3. Butterfly valve, wafer/flangeless, with replaceable body seat

Fig. 4. Butterfly valve, wafer U-section rubber seated, with gear operating unit

CHECK VALVE

Check valves are also described as reflux, non-return, back pressure, retaining and clack valves. In some areas of usage the term 'reflux' is reserved for the swing disk type of valve and the description 'check' is largely associated with the lifting disk type of valve. Current British Standards use the term 'check valves' and this includes both swing and lift types.

The purpose of a check valve is to permit flow in one direction and to prevent it in the reverse direction. Commonly such valves are automatic in operation, the pressure of the fluid flowing in one direction holding the valve disk open and reverse flow, together with the weight of the disk, acting to seat the disk and so cut off the flow.

TYPES

Most check valves are based on either the swing concept or the lift concept. There are many variants and some of the more commonly used of these are referred to in the following. A brief description is also given of a new concept in this field, the cone and diaphragm check valve.

Swing Check (conventional)

In this design (Fig. 1) a hinged, gatelike disk is free to swing against a body seating face which is tilted. To ensure that the disk takes up a satisfactory seating position at all times, the disk to hinge connection is designed so that the disk has sufficient freedom to swivel and make true and full contact with the body seat. Disks may be all metal or may be fitted with leather, rubber, or composition facings, depending on the nature of the service.

In the fully open position there is little obstruction to flow in a swing type check valve and consequently the pressure drop across the valve is relatively small.

Swing Check with Outside Lever and Weight

In this arrangement (Fig. 2), the disk hinge pin is extended through the body and fitted with an outside lever and weight, which can be set in various positions.

In the position illustrated the weight will assist the disk to close quickly once forward flow ceases. Quick closing action is desirable in systems where sudden reversals of flow may occur and accumulate considerable momentum before an ordinary unweighted valve disk would close. Quick closing minimizes possible damaging shock and disk chatter.

The lever and weight may be mounted so that the weight balances or partly balances the weight of the disk. The valve will then be more sensitive to low pressures and velocities.

Swing Check with Outside Lever and Quadrant

(Fig. 3) The lever has a drilled hole and a locking pin fastens the lever to a quadrant, thereby holding the valve disk in the open position. Used mainly to assist in line pressure testing.

Swing Check with Fusible Link

(Fig. 4) The disk is held in the open position by means of a fusible link attached to the lever. The fusible link has a predetermined melting point and the valve is used mainly as a fire control device.

Swing Check with Shock-free Operation

This is a type of swing valve (Fig. 5) in which the design criteria are directed particularly towards the prevention of shock closure in conditions of very rapid reversal of flow, for example in the case of discharge by electrically driven centrifugal pumps. Special features are incorporated to achieve valve closure in the shortest possible time combined with low head loss.

Tilting Disk Check

A sophisticated design (Fig. 6) of swing check valve in which a specially shaped disk is pivoted at a selected point, instead of being hinged as in the conventional type of swing check valve. The seat faces on the disk and in the body are bevelled and in the closed position the valve resembles a simple 'lift' valve with conical seat.

Characteristics are low pressure drop at low velocities and, depending on the design, a quick response to flow reversal.

Dual Plate Check

(Fig. 7) The valve body is cylindrical in shape and the disk is in the form of two semicircular plates attached to a central hinge pin located in the body. The disk plates are acted upon by one or more torsion springs mounted on the hinge and these hold the plates against a flat seating in the body. The pressure of reverse flow on the plates causes the springs to deflect and allow the plates to move to an open position.

Each plate may be acted upon by a spring or springs independently of the other plate, and on larger valves each plate may be independently supported. The incorporation of such features improves plate response when rapid changes of flow rates are experienced and reduces the severity of pressure surges.

Lift Check

(Fig. 8) The disk sits on a seat face provided on a horizontal bridge wall across the valve body, rather as in a globe valve except that in this case the disk is free to rise. Flow pressure lifts the disk from its seat and back flow, or no flow, causes the disk to drop back on to the seat and so shut off the flow.

The disk is usually guided in both the body seat opening and the body cap but sometimes within the seat opening only. For some applications a spring may be fitted

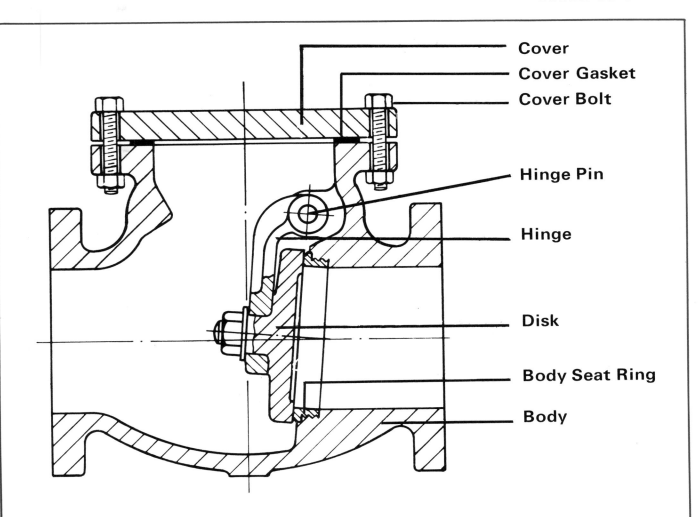

Cover
Cover Gasket
Cover Bolt
Hinge Pin
Hinge
Disk
Body Seat Ring
Body

Fig. 1. Typical swing check valve

Fig. 2. Swing check valve with outside lever and weight

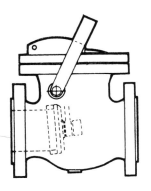

Fig. 3. Swing check valve with outside lever and quadrant

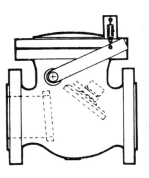

Fig. 4. Swing check valve with fusible link for holding open

above the disk to ensure that the valve is normally closed.

Depending on the service conditions, disks may be of all metal construction or may be in the form of a disk holder fitted with a rubber or composition disk ring.

Like the globe valve the flow path through a lift check valve follows a twisting course. Consequently the pressure drop across the valve is greater than in the swing type of check valve where the flow is less restricted.

Lift Check with Dashpot

In this design (Fig. 9) a dashpot is arranged above the disk so as to provide a cushion effect on services where pulsations in the line may cause an ordinary check valve to hammer.

Ball Check

Similar to the normal lift type check valve but utilizing a ball as the disk. The ball is guided in the body cap but free to rise and rotate.

Screw-down Stop and Check

Also known as a screw-down non-return (SDNR) valve (Fig. 10). This is a globe type of valve except that the disk is not attached to the screwed stem. The stem is used to regulate the lift of the disk when it is acting as a check valve, or, to hold the disk closed in the same manner as an ordinary globe valve.

Cone and Diaphragm Check

This is a modern development (Fig. 11) and a new concept in the field of check valves. The essential elements are a stainless steel perforated cone and a flexible diaphragm of natural or synthetic rubber. Service fluid passes through holes in the cone and deflects the rubber diaphragm inwards. As the flow increases the diaphragm is progressively deflected inwards until the valve is fully open. When flow is reduced the diaphragm tends to open outwards to its original shape and when flow ceases the diaphragm snaps shut, closing the valve.

There are no metallic moving parts in the valve, which is extremely lightweight and can be installed in horizontal or vertical pipes even where the flow is vertically downwards. Operating temperatures and pressures are limited to the capabilities of the diaphragm material.

Sizing of Check Valves

Care should be taken to avoid oversizing of conventional swing and lift check valves. To obtain the minimum velocity required to lift the disk to the full open and stable position it may be necessary on some applications to fit valves smaller in size than the pipe in which they are installed.

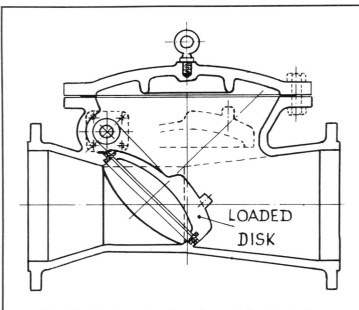

Fig. 5. Swing check valve with shock free operation

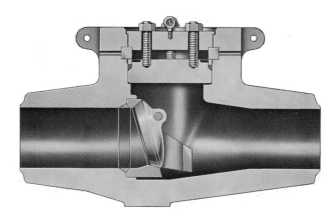

Fig. 6. Tilting disk check valve with pressure sealed top closure

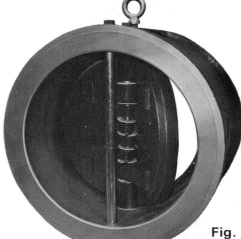

Fig. 7. Dual plate check valve

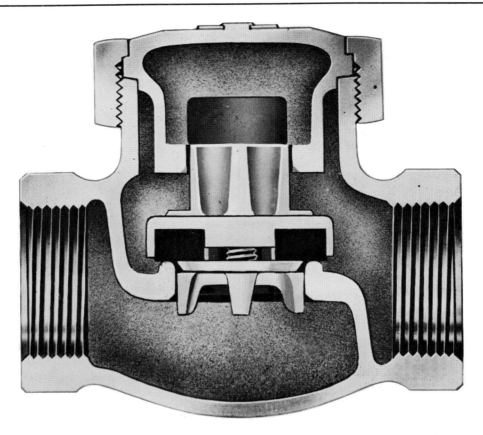

Fig. 8. Typical lift check valve (with composition disk seating ring)

Fig. 9. Lift check valve with dashpot

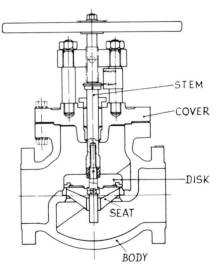

STEM

COVER

DISK

SEAT

BODY

Fig. 10. Screw-down stop and check valve

Fig. 11. Cone and diaphragm check valve

DIAPHRAGM VALVE

A general description of this valve is one in which a flexible diaphragm is, or is associated with, the closure member and in which the diaphragm isolates the valve operating mechanism (bonnet) from the fluid passageway of the valve. A typical design of diaphragm valve is illustrated in Fig. 1.

The separation of operating mechanism from pipeline content guarantees the purity of the line fluid and conversely prevents any possibility of attack by the line fluid on the working parts. In addition, the need for any form of separate sealing for the stem is eliminated, except as a safeguard on applications where dangerous fluids are being handled.

As the fluid contacts only the diaphragm and the body, both of which are available in many different materials, the valve is ideally suited for controlling a wide range of fluids, especially the chemically aggressive or abrasive.

Operating temperatures of diaphragm valves are generally limited by the material used for the diaphragm or body linings. Depending on this, temperature conditions from around − 50 °C to 175 °C can be accommodated.

The simplicity of the construction of the diaphragm valve, in which there are only three main parts – body, diaphragm, and bonnet assembly – permits easy and quick dismantling for maintenance. Should a diaphragm require changing, this can be done in-situ and in a relatively short time.

Separation of operating mechanism from pipeline contents makes the diaphragm valve ideally suited for use with edible and medical products as well as 'hard to handle' and dangerous fluids. More generally, the many grades of elastomers and plastics available for diaphragms and the wide choice of materials for bodies and linings have enabled the diaphragm valve to find acceptance and application in practically every aspect of modern industry.

TYPES

Although there are a number of design variants the two main forms of diaphragm valves in use today are the weir and the straight-through types.

Weir
The weir type (Fig. 1) is the most widely used; tight shut-off is obtained with comparatively low operating force and short diaphragm movement, which minimizes the amount of flexing required of the diaphragm, so lengthening diaphragm life and reducing maintenance, downtime, and costs.

The diaphragm is made either of an elastomer or of PTFE with an elastomeric backing, and is connected to a compressor component which is attached to the threaded stem. To close the valve, the diaphragm is pressed down to make a tight seal against a weir formed in the body, or the contour of the body or a portway in the body, depending upon the particular design of valve.

Straight-Through
Figure 2 shows a typical diaphragm valve of the straight through flow type. There is no weir, so flow through the valve is in a straight line along a full bore passageway. This feature makes the valve particularly well suited to handling viscous fluids, thick slurries, and fluids containing deposits. The stroke of the diaphragm is appreciably longer than in the weir type of valve and this limits the choice of materials for the diaphragm to elastomers.

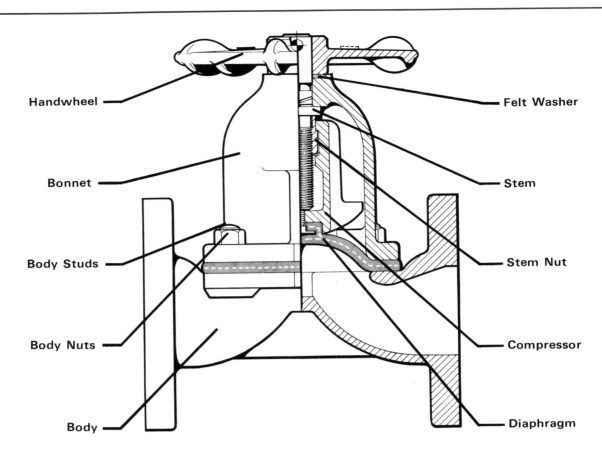

Handwheel

Bonnet

Body Studs

Body Nuts

Body

Felt Washer

Stem

Stem Nut

Compressor

Diaphragm

Fig. 1. Weir type diaphragm valve

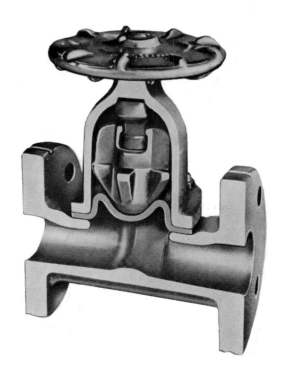

Fig. 2. Diaphragm valve with straight through flow

GATE VALVE

Of all the many different types of valves available to industry the gate valve is by far the most widely used. In this valve a gate-like disk is moved at right angles across the line of flow between matching seats in the valve body, thus opening or shutting off the flow. It is intended for duty as a stop valve and in the fully open position provides straight through full bore flow with consequent minimum loss of pressure of the service fluid.

Conventional types of gate valves are generally ideal for services that require infrequent operation and where the disk is kept fully open or fully closed. They are not intended for modulating or throttling purposes. Velocity of flow against a partly opened disk may cause vibration and chattering with possible damage to the seating surfaces and throttling can subject the disk to the erosive effects of wire drawing.

TYPES

The principal variation in the design of gate valves is in the type of sealing element employed and reasonably the valves are often described in terms of the sealing element, e.g., solid wedge, parallel slide, parallel double disk, and split wedge. There are many other types, some of which are included in the examples shown on pages 25 and 27, but the designs in most common use are the wedge gate valve and parallel slide valve.

Wedge Gate Valve
The fluid controlling element in this design of gate valve (Fig. 1) is a rigid one-piece, wedge shaped, disk containing no loose parts. The disk is accurately guided in the valve body to ensure that contact between the disk seat faces and body seat faces is limited to only a small portion of the closing and opening movements. Because of the wedge action, tight sealing can be achieved without assistance from the fluid pressure.

Parallel Slide Gate Valve
The design feature of this valve (Fig. 2) is a parallel and flexible sealing element containing two half-disks which slide between two corresponding parallel seat faces in the body. A spring is usually contained between the two halves of the disk. This spring has three functions: to provide a flexible closure member, to hold the disk halves in sliding contact with the body seats under zero or near zero pressure so that a wiping action is obtained, and to dampen down any tendency to vibrate of the disk parts.

Closure sealing is achieved by the thrust of the line pressure forcing the disk against the seating face on the downstream side of the body and the resultant contact pressure secures fluid tightness. The closure member of a parallel slide valve has a high degree of flexibility, which is particularly useful on services subject to appreciable variations of temperature. Expansions or contractions of working parts can be accommodated without affecting the operation of the valve.

Parallel Double Disk Gate Valve
This valve (Fig. 3) has two parallel disk halves which are forced outwards against the body seat faces by means of a spreader or wedge when contact is made with a stop in the bottom of the valve body. The first opening movement releases the disk halves and continued operation raises them clear of the body seat openings.

Split Wedge Gate Valve
Similar to the parallel double disk valve but the disk is wedge shaped (Fig. 4) Again a spreader device is used to force the disk halves against matching tapered seats in the valve body and the first opening movement releases the disk halves from contact with the body seats

DESIGN FEATURES

Flexible Wedge Disk
The wedge type disk can also be obtained in a flexible design, developed primarily to overcome sticking on high temperature services subject to great temperature changes. In one form the single piece construction of the disk is maintained but the two seating faces are separate from each other except for a short connecting axle or spud at the centre of the disk (Fig. 5).

In this way each disk face is permitted a certain degree of independent movement which, besides eliminating the possibility of sticking in the closed position, facilitates tightness on both the upstream and downstream sides of the disk over a wide range of pressures and temperatures.

Soft Sealing
Soft sealing is a recent development in which plastic or synthetic rubber rings are inserted into the faces of the disk or body seat rings to provide soft sealing backed up by contact of the metal to metal faces. Soft seals provide improved sealing characteristics over a very useful range of temperatures for a great variety of fluids. The resilience of the seal gives good recovery from deformation by accumulated 'dirt' and solid particles in the service fluid. Another asset is the reduction in effort required to operate the valve.

For block and bleed control, sometimes known as double blocking, the use of soft seals can secure the requirement for positive tight closure on both sides of the disk with the aid of only one valve. This may avoid the possible use of two conventional valves to obtain the same results.

Another development in soft sealing offering positive tight shut-off with high resilience is the wedge gate valve

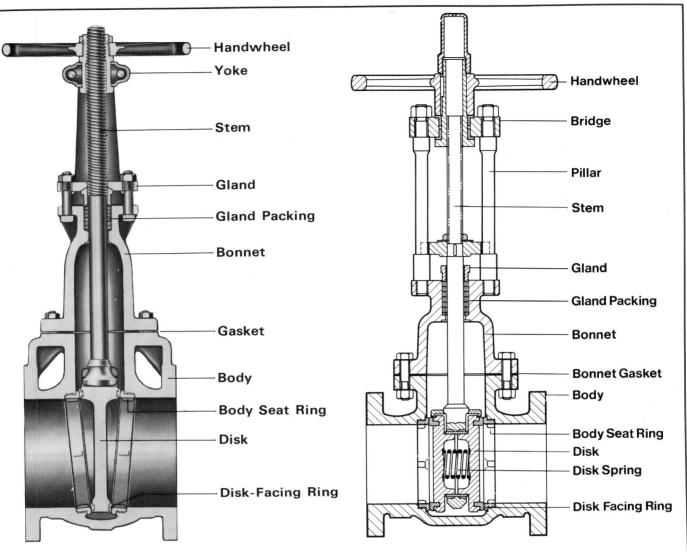

Handwheel

Yoke

Stem

Gland

Gland Packing

Bonnet

Gasket

Body

Body Seat Ring

Disk

Disk-Facing Ring

Fig. 1. Typical solid wedge gate valve with outside screw and rising stem

Handwheel

Bridge

Pillar

Stem

Gland

Gland Packing

Bonnet

Bonnet Gasket

Body

Body Seat Ring

Disk

Disk Spring

Disk Facing Ring

Fig. 2. Parallel slide gate valve

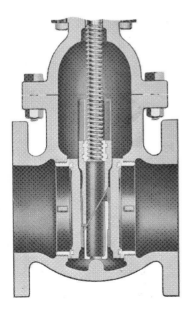

Fig. 3. Parallel double disk gate valve

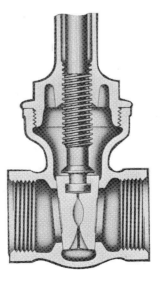

Fig. 4. Split wedge gate valve

Fig. 5. Flexible wedge disk

in which a facing of synthetic rubber is bonded to the disk casting. Both upstream and downstream faces and also the lower periphery of the disk are covered with the rubber. Sealing round the upper two-thirds of the disk is axial by wedge action while round the lower one-third of the disk the seal is achieved by radial contact between the rubber covered lower part of the disk and the base of the body.

Stem Arrangement

There are three basic designs of stem arrangements:
(1) Inside screw, rising stem (Fig. 6).
(2) Inside screw, non-rising stem (Fig. 7).
(3) Outside screw, rising stem (Fig. 8).

The first mentioned is relatively the least expensive and is most commonly found on the smaller sizes of valves. A useful feature is that the position of the stem serves to indicate the position of the disk. Because the stem threads are inside the valve body and so open to attack by the service fluid, inside screw valves are not usually used for fluids having corrosive or erosive properties or for high temperature services where consequent expansion and contraction may cause binding of the threads.

In the case of the inside screw, non-rising stem design, the stem does not move axially but merely rotates. This arrangement is particularly useful where headroom is limited. Also, the elimination of the up and down movement of the stem reduces the amount of wear on the gland packing.

On the outside screw, rising stem valve, the stem threads are situated outside the valve body and so are not subjected to possible effects of the pipeline fluid. The stem threads are accessible for lubrication and the position of the stem provides an indication of the amount of valve opening. Adequate headroom is required for the rising stem, for which some form of protection should be arranged to guard against possible damage.

Fig. 6. Inside screw, rising stem

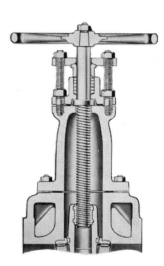

Fig. 7. Inside screw, non-rising stem

Fig. 8. Outside screw, rising stem

OTHER GATE VALVE DESIGNS

Lever or Quick Acting

In one design (Fig. 9) of quick acting gate valve a lever takes the place of the handwheel and operates a sliding stem. In another design (Fig. 10) a lever is connected to the disk by a shaft and link arrangement and a quarter-turn rotation of the lever opens or closes the valve.

Clamp or Clip

The body and bonnet are held together by means of a single 'U' clamp (Fig. 11). This arrangement is ideal where frequent inspection and cleaning of lines are necessary: by removing two nuts the complete bonnet assembly is easily removed. The slimness of the valve reduces space requirements and cuts down on weight.

Fig. 10. Quick acting type of gate valve — shaft and link (quarter turn)

Fig. 9. Quick acting type of gate valve — sliding stem

Fig. 11. Clamp gate valve

TYPES OF VALVES

Conduit

This type of valve (Fig. 12) employs a parallel faced disk that is lengthened to include a circular part of the same diameter as the valve bore. The valve is characterized by its body design, which is extended above and below the centre line to provide the necessary cavities for the virtually double length disk. A full bore smooth and continuous flow path is obtained and the use of product separators is made convenient.

Penstock

A single faced gate or door is moved vertically or horizontally between guides attached to a frame (Fig. 13) that is fixed to a wall or bulkhead. The guides may be parallel to the frame, in which case sealing is in one direction only and dependent on the fluid pressure, or they may be tapered to provide a wedge effect between frame and guides, thereby obtaining gate tightness in both directions.

It is used for handling large volumes of water in waterworks and sewage schemes, also as a watertight door on ships.

Line-Blind, Spectacle, or Goggle

A spectacle shaped plate with one end blank and an orifice in the other end is rotated about its centre between two flanges (Fig. 14). When the flange bolts are tightened the spectacle plate is sealed between the flanges in either the open or the closed position. In another design the spectacle or goggle plate is raised and lowered by a threaded stem and handwheel or other means. As one end of the spectacle plate is always visible outside the valve body this provides an indication of whether the valve is open or closed.

Pulp Stock, Knife, or Plate

A valve of rugged construction to meet the requirements of pulp stock handling. Usually has a plate disk with a knife-like cutting edge at the bottom (Fig. 15) to assure complete shearing and positive shutoff.

Fig. 12. Conduit valve

Fig. 13. Penstock valve

Fig. 14. Line-Blind valve

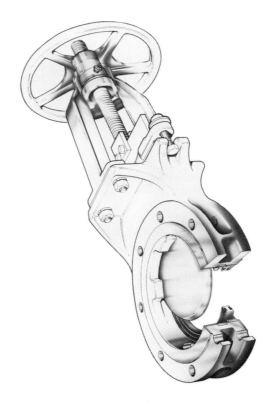

Fig. 15. Pulp stock or knife valve

GLOBE VALVE

In this screw-down stop valve, or globe valve as it is commonly called, the axis of the stem is at right angles to the body seat face. The relatively short movement required of the stem to open or close the valve and the very positive seating action combine to make this type of valve well suited for duties as a stop valve or for the close regulation or throttling of flow.

All contact between body seat and disk ends as soon as the disk is moved from the closed position so that mechanical wear of the sealing faces is minimized. Whatever wear may occur during service does not create a major problem as the body seat and disk in most globe valves can be repaired or replaced with a minimum of difficulty and without having to remove the valve from the pipe line, a distinct advantage where welded-in valves are concerned.

The minimum resistance to flow of a globe valve is higher than in most other valves because of the changes in direction of flow as the fluid passes through the valve. This may vary considerably, however, depending on the body design and the disposition of the stem relative to the inlet and outlet ports.

STEM ARRANGEMENT

In common with gate valves, both inside and outside screw arrangements are used for globe valves. The larger valves are usually of outside screw design, while the inside screw arrangement predominates for the smaller ones.

SEATINGS

Various seating designs are available to suit different service requirements and provide different flow characteristics.

The conventional narrow seat type (Fig. 1) has a tapered or spherical shaped surface on the disk and a flat angled surface on the body seat and there is narrow line contact between disk and seat. Provided suitable materials are used for disk and seat, this line bearing breaks down hard deposits that may form on the seat on some services and assures pressure-tight closure.

There is little protrusion of the disk into the valve orifice so that full bore flow is obtained with only a short lift of the disk, which makes this form of seating unsuitable for close regulation of flow. Preferably, such a valve should be used wide open or fully closed.

To achieve better flow characteristics the disk is made to project into the body seating orifice. The disk may be of the plug type (Fig. 2), which is conical in shape with matching conical seat in the body, giving reasonable flow control and exceptional resistance to galling, erosion, and wire drawing under throttling conditions, or it may be contoured to provide specific flow characteristics, e.g., percentage flow equal to percentage lift. Other specialized designs may use hollow disks with vee shaped or contoured body seatings to achieve a variety of flow conditions.

'Soft' forms of seating are also employed, including PTFE (or other plastic) seat or disk inserts and the very popular composition disk (Fig. 3), which is used frequently for steam and gas services, particularly in low pressure bronze valves. 'Soft' seatings provide tight shut-off with the minimum of effort but they are not suitable for throttling duties as they can be quickly damaged by wire drawing. Disk replacement is a simple matter and, providing the body seat face is undamaged, seating performance can be quickly restored to 'as new' condition. Also, by using disks of different materials a valve may be made suitable for other classes of service.

Generally the disk and stem are separate components and connected together in such a manner that the disk is free to revolve independently of the stem and is able to swivel. This allows the disk to sit squarely on its seat and avoids frictional contact that might damage the seating surfaces. In some small valves, e.g., needle valves, the disk and stem may be integral, and while these will give close and stable regulation of flow under high pressure drop conditions, they are not best suited for use as shut-off valves.

Although much larger sizes are made, standard lines of globe valves do not usually extend beyond 200 mm or 250 mm. Beyond this, depending on the service pressure, the axial load imposed on the stem by the pressure of the fluid acting over the whole of the exposed area of the disk can make direct manual operation very difficult or even impossible. To overcome this problem some form of additional mechanical advantage can be provided, such as hammer-blow wheels, gearing, or power operators. Special designs are also available which incorporate means for balancing or partially balancing the axial load on the disk.

TYPES

Angle Valve

In the angle type of globe valve (Fig. 4) the flow has to make only one change of direction so that the pressure drop across the valve is much less than in a conventional design of globe valve. When installed to the best advantage angle valves can reduce the number of fittings in a pipe system by serving as a valve and a 90° elbow.

Oblique Valve

In the oblique or 'Y' type globe valve (Fig. 5) the orifice in the body is arranged at an incline to the normal flow line. This results in less disruption of the flow pattern than in a conventional globe valve and a correspondingly smaller loss of pressure across the valve.

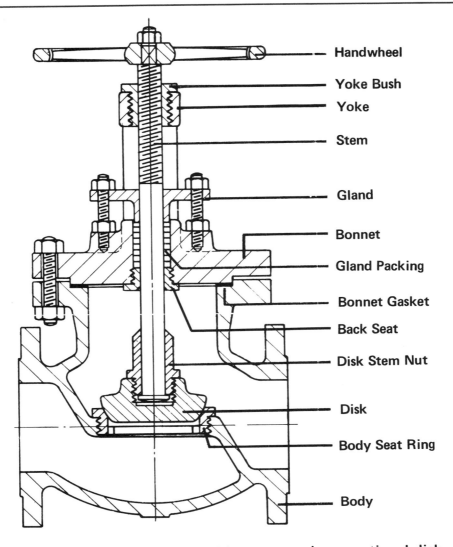

Handwheel
Yoke Bush
Yoke
Stem
Gland
Bonnet
Gland Packing
Bonnet Gasket
Back Seat
Disk Stem Nut
Disk
Body Seat Ring
Body

Fig. 1. Globe valve with outside screw and conventional disk

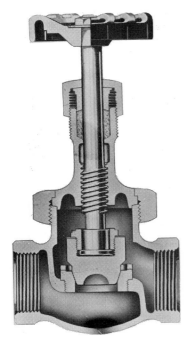

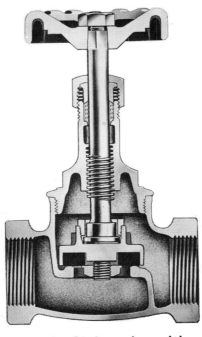

Fig. 2. Globe valve with inside screw and plug disk

Fig. 3. Globe valve with inside screw and composition disk

Fig. 4. Angle globe valve

TYPES OF VALVES

Needle Valve

The needle valve (Fig. 6) is designed to provide fine control of flow and is generally restricted to small sizes of pipes. The disk, commonly integral with the stem, has a needle shaped end which fits very accurately into the body seat and the threads on the stem are made of finer pitch than usual to obtain close regulation of flow. Usually needle valves have a reduced size of orifice in relation to the pipe size.

Piston Type

This type (Fig. 7) is a variant of the conventional globe valve in which the usual form of disk and seating is replaced by a design of seating based on the piston principle. A precision ground piston is connected to the stem and the seal is formed by two resilient sealing rings surrounding the piston. The rings are separated by a lantern ring and are compressed firmly around the piston by the load exerted on the cover by the cover nuts.

The resilient rings are replaceable and can be provided in various materials including PTFE.

The valve is intended primarily for 'open' and 'shut' duties but can be fitted with a specially shaped piston or special lantern ring for close regulation of flow.

Fig. 5. Oblique or Y type globe valve

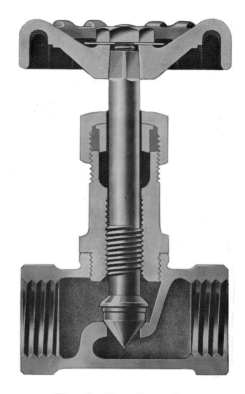

Fig. 6. Needle valve

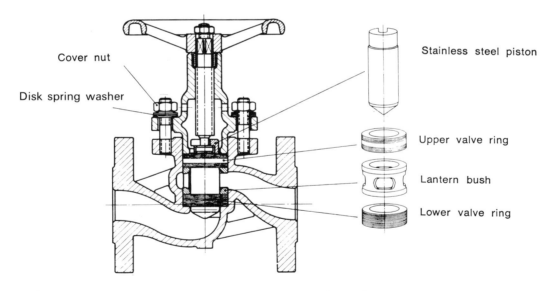

Cover nut

Disk spring washer

Stainless steel piston

Upper valve ring

Lantern bush

Lower valve ring

Fig. 7. Piston type globe valve

PLUG VALVE

The plug valve is a development of the simple cock. It has a plug, tapered, or cylindrical, which can be rotated to move its ports relative to the body ports, thus controlling the fluid flow. Construction is simple and compact with few basic components. The quarter-turn operation from fully open to fully closed provides quick action, and flow is straight through the valve with no sudden alterations in shape or section, so that loss of pressure head due to the valve is low.

An important characteristic of the valve is the ease with which it can be adapted to multi-port construction, so that one valve will provide two, three, or even four different flow ways. This can simplify the piping layout and reduce the number of valves and fittings required in an installation.

The primary service of a plug valve is positive open–close operation but they can be used for coarse throttling on some low flow services.

TYPES

Plug valves may be divided into two categories, non-lubricated and lubricated.

Non-lubricated sleeved plug valves are used in the chemical and petrochemical industries and have many other service applications, particularly on installations where lubricants are unacceptable.

Lubricated plug valves are used extensively in oilfield production, distribution, and refinery installations. They are also widely used in the petrochemical and heavy chemical industries, in the gas distribution, heating, and ventilating industries, and for general services.

Non-lubricated Valves

The basic example is the simple plug cock. An adjustable gland may be incorporated and in some designs the plug may be spring loaded to provide a means of compensating for wear, to assure tightness, and to permit easy operation.

A significant advance has been the introduction of the sleeved plug valve in which a 'soft' sleeve is arranged between the plug and the body (Figs 1 and 2). The sleeve material may be one of the fluorocarbon group of plastics, such as PTFE, or some kind of composition, depending on the service potential of the valve. The whole flow passage can also be lined with the same fluorocarbon material.

Sleeved plug valves, with tapered or parallel plugs, assure stick-free operation with improved sealing and require less maintenance. Additionally, the self lubricating properties of plastic materials of the PTFE type make the valve particularly suitable for applications where conventional valve lubricant is unacceptable.

The pressure–temperature range of sleeved valves is limited by the seat material used and with the present fluorocarbons is about 29 bar and 220 °C to 230 °C.

A variant of the non-lubricated plug valve is shown in Fig. 3. In this design the plug takes the form of two segments that are pressed against the body by the action of a central wedge piece. This wedge is attached to the valve stem and contains the flow way through the plug. Instead of a body sleeve the design incorporates the feature of preformed PTFE seal inserts in the plug segments.

Lubricated Valves

These are made with either tapered or parallel plugs (see Figs 4 and 5). An insoluble lubricant/sealant is injected under pressure to form a film between the plug and body surfaces. The special lubricant or compound is fed into the operating stem of the valve by means of a pressure screw or grease gun, passes through a non-return valve, and reaches the seating surfaces via a system of ducts and grooves in the plug and body.

As well as facilitating valve operation, the special lubricant also perfects the seal between the accurately matched seating surfaces and assures positive leak tightness. Additionally it serves to protect the seating surfaces from corrosion and erosion. To overcome any initial resistance to operation of tapered plug valves after a long period of inactivity, the lubricant pressure can be built up in the chamber at the small end of the plug so that the plug is lifted slightly in its seat and easy operation restored.

Lubricant/sealants have been greatly developed in recent years and are now available to meet the requirements of nearly every type of service. Plugs and body seating surfaces can also be coated with a fluorinated plastic, such as PTFE, to increase the lubricosity and hence the valve performance.

Depending upon materials and design, lubricated plug valves are available for pressures up to 690 bar. The lubricant/sealment used limits the temperature range to approximately −40 °C to 325 °C.

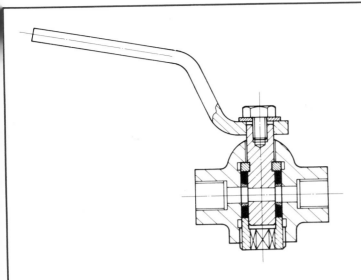

Fig. 1. Sleeved plug valve with parallel plug

Fig. 2. Sleeved plug valve with tapered plug

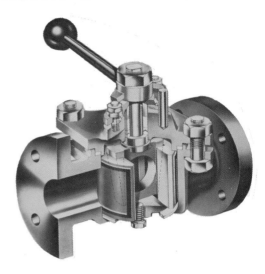

Fig. 3. Split non-lubricated plug valve with preformed PTFE seal inserts in parallel segmented plug

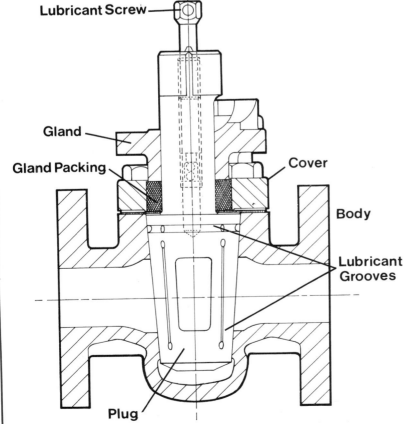

Lubricant Screw

Gland

Gland Packing

Cover

Body

Lubricant Grooves

Plug

Fig. 4. Lubricated plug valve with tapered plug

Fig. 5. Lubricated plug valve with parallel plug

PRESSURE CONTROL — REDUCING AND RETAINING VALVES

The reference here is to self operated valves used to control pressure in a system and which do not require an external power source. When valves equipped with power actuators are used for this purpose they come within the field of 'automatic process control valves.'

PRESSURE REDUCING VALVE

This is installed where it is required to reduce from one level of pressure to another and to maintain the reduced pressure on the downstream side within limits, irrespective of fluctuations in the inlet pressure or change in flow demand. The valve is automatic in operation.

PRESSURE RETAINING VALVE

Also known as a surplus valve and used to maintain a level of pressure in the line upstream of the valve, the valve opening with rising upstream pressure. It is usually a reverse acting version of the pressure reducing valve.

TYPES

Self operated pressure reducing or retaining valves fall into two main categories, direct acting and pilot operated.

Direct Acting Valve
The controlled pressure acts directly through a diaphragm, piston, or bellows, on an imposing force from a compressed helical spring, weight, or weighted lever, or from compressed air. The construction is simple and robust and such a valve can provide long life with maintenance free operation even under adverse working conditions.

Although the pressure control provided by direct acting valves is not so accurate as with pilot operated valves they are less costly and have many applications for which the fine control offered by the latter would be unnecessary. A typical example of a direct acting valve is shown in Fig. 1.

Pilot Operated Valve
The main valve is either assisted or completely controlled by the operation of a pilot valve, which may be itself a small direct acting reducing valve.

The precise method of operation depends on the particular design of valve, but essentially the pilot valve acts so as to regulate the amount of opening of the main valve in a way that will maintain the flow at the desired level of pressure.

Pilot operated valves provide very close accuracy of pressure control, are compact in design, and are usually much smaller than direct acting valves for the same duty.

The pilot valve may be integral with the main valve or may be a separate unit suitable for remote pressure sensing. It can also be used for remote on–off control, i.e., as part of a complex system governed by a central control. Further, the facility of direct control by temperature can be provided by fitting the appropriate type of pilot valve.

Because of the complexity of design, pilot operated valves require regular maintenance and clean working conditions, the latter often being ensured by the fitting of a strainer immediately upstream of the valve.

Figure 2 illustrates a typical pilot operated valve of the diaphragm and piston design.

Valves can be designed with a single seat, which usually indicates the ability to close tight under 'no flow' or 'dead end' conditions, or they can be of double seat construction, which improves the maximum flow rate and accuracy of pressure control but incurs the penalty of the loss of ability to control pressure at zero or very low flow rates.

Duties vary over a wide field of applications, including steam, compressed air, industrial gases, water, oil, and many other liquids. Consequently, and in view of the many possible design variations, it is essential when considering selection that the exact duty to be performed is first fully explored to ensure satisfactory operation. Full and comprehensive information should be supplied to manufacturers whose specialized assistance is readily available.

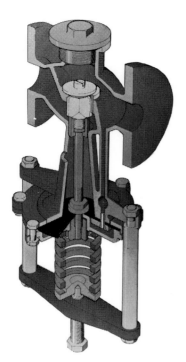

Fig. 1. Direct acting pressure reducing valve

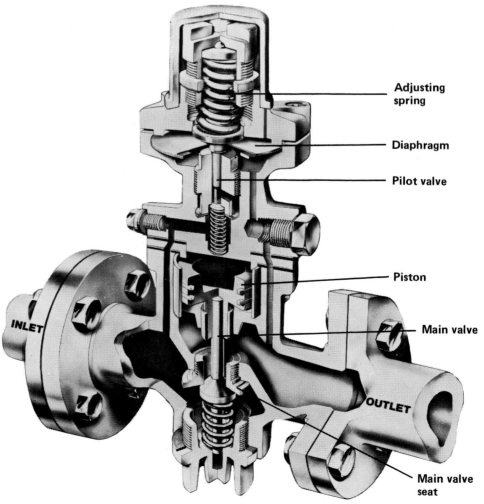

Adjusting spring

Diaphragm

Pilot valve

Piston

Main valve

Main valve seat

INLET

OUTLET

Fig. 2. Pilot operated pressure reducing valve

SAFETY, RELIEF, AND SAFETY RELIEF VALVES

These are valves for automatically preventing a safe pressure being exceeded in fired or unfired pressure vessels or pipeline systems.

BASIC TYPES

Safety Valve
A valve (Fig. 1) which automatically discharges steam, gases, or vapours so as to prevent a predetermined safe pressure being exceeded. Such valves usually have a rapid opening action (pop action), and obtain their rated discharge capacity with a rise in pressure (over-pressure) of 10 per cent or less. The reset pressure is also closely controlled.

Relief Valve
A valve (Fig. 2) which automatically discharges liquid so as to prevent a predetermined safe pressure being exceeded. The term is commonly used for pressure relieving valves in which the lift is proportional to the increase in pressure above the set pressure.

Safety Relief Valve
A valve (Fig. 3) which, depending on its application, automatically discharges gases, vapours, or liquids so as to prevent a predetermined safe pressure being exceeded.

Safety or relief valves should be used on any closed vessel or system in which the pressure can be other than atmospheric and where under any circumstances the design pressure of the system can be exceeded. In any process, imbalance in rates of fluid flow or energy transfer into or out of process equipment may result in the pressure exceeding the operating pressure. If, for these or other reasons, the pressure exceeds prescribed limits it must be relieved by a safety or relief valve.

The reliability of such valves is of special importance for the protection of the operator and the plant itself. They are the subject of mandatory regulations and insurance acceptance in most countries and standards exist dealing with their design and use. Careful calculation based on established practice and known data is required in order to establish the correct size and type of valve, not only to suit the working conditions but also to ensure that the capacity of the valve selected is sufficient to protect the system.

A high proportion of safety or relief valves discharge the surplus pressure direct to atmosphere, but in the case of 'difficult' fluids which are toxic, corrosive, or flammable, discharge may be via a complex disposal system. It is not uncommon for safety or relief valves in such installations to be connected into a common disposal system, and this requires valves which are designed for that purpose.

SIZING AND CODES OF PRACTICE

Sizing of safety or relief valves is of the utmost importance in order to guarantee that the discharge is adequate to protect the system fully. In order to ensure this, sizing must be in accordance with the code of practice used in the design of the pressure vessel or pipeline system.

MATERIAL SELECTION

Careful consideration has to be given to the selection of materials of construction, taking account of the requirements for both non-relieving and relieving conditions.

BASIC CLASSES

There are essentially three basic classes of safety or relief valves, to which special design features may be added to give particular operational characteristics:
1. Direct (Figs. 1, 2, 3, 5 and 6)
2. Pilot Operated (Fig. 4)
3. Supplementary Loading.

Direct Acting Valve
The direct acting valve is the simplest and most commonly used class because it is suitable for most applications. The load is usually applied to such valves by means of helical coil compression springs, although other means of loading are sometimes used, such as weights.

When valves are required for use on 'difficult' fluids which may be odorous, toxic, corrosive, or flammable, a totally enclosed valve which does not permit any discharge to atmosphere, other than through the normal outlet, must be used. It is sometimes permissible, however, for valves to vent to atmosphere other than through the normal outlet when used with steam or inert gases (e.g., open bonnet valves – see Fig. 5). If a valve or group of valves relieve into a common discharge system which can result in a variation in back pressure on the valve outlet, consideration should be given to the fitting of a balancing device such as a bellows (Fig. 6) in order to eliminate the effects of the back pressure on the set pressure of the valve.

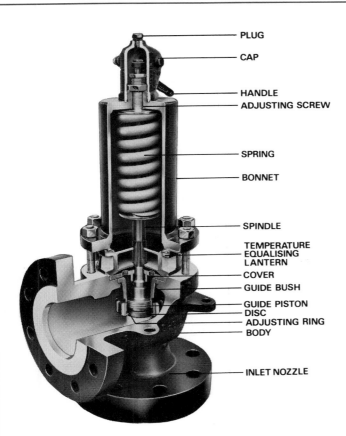

PLUG
CAP
HANDLE
ADJUSTING SCREW
SPRING
BONNET
SPINDLE
TEMPERATURE
EQUALISING
LANTERN
COVER
GUIDE BUSH
GUIDE PISTON
DISC
ADJUSTING RING
BODY
INLET NOZZLE

Fig. 1. Typical safety valve

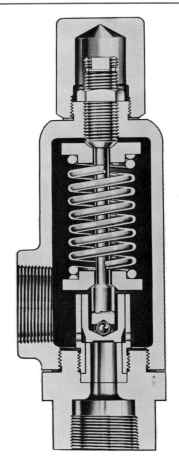

Fig. 2. Proportional
lift relief valve

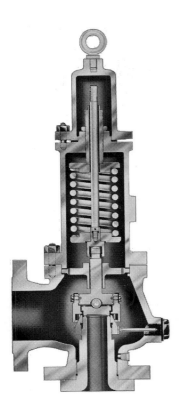

Fig. 3. Safety relief
valve

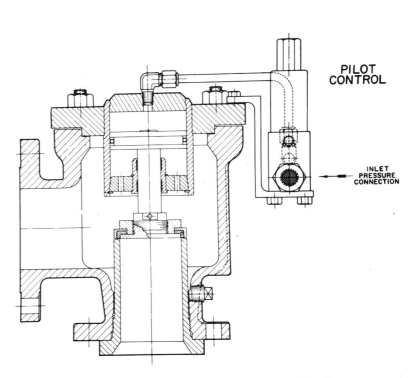

PILOT
CONTROL

INLET
PRESSURE
CONNECTION

Fig. 4. Pilot operated safety relief valve

TYPES OF VALVES

Pilot Operated Valve

This takes several forms, but the most common type comprise a main valve and a pilot valve through which the system pressure loads the main valve. When a predetermined pressure (set pressure) is reached the pilot valve, which is itself a safety or relief valve, relieves the pressure loading on the main valve and allows it to open. Pilot operated valves are used when the increase in pressure to open the valve, and the fall in pressure to allow the valve to close (blowdown), must be less than can be attained by the use of direct acting valves.

Supplementary Loading Valve

This is similar to the direct acting valve, except that an additional load is applied from an external source.

Electric and pneumatic supplementary loading systems are available. These are used where it is necessary for the system pressure to operate closer to the relieving pressure of the valve than can be obtained with a direct acting valve, while retaining the fail-safe features of the latter.

Because of the many factors to be considered in selecting the correct valve, specialist manufacturers should be consulted. For further information on safety or relief valves and associated relief system design reference may be made to relevant publications by the British Standards Institution, American Petroleum Institute, American Society of Mechanical Engineers, TUV Merkblatt, etc.

GLOSSARY OF SOME TERMS IN COMMON USE

Set Pressure The pressure measured at the valve inlet at which a safety valve commences to lift under service conditions.

Blowdown Otherwise known as pressure drop of a safety valve, it is the difference between the set pressure and the reseating pressure expressed as a percentage of the set pressure or as a pressure difference.

Back Pressure The pressure at the outlet of the safety valve.

Superimposed Back Pressure The pressure at the outlet of a safety valve, connected to a discharge system, before the valve opens. (*See* built-up back pressure.)

Built-up Back Pressure The pressure existing at he outlet of a safety valve caused by flow through the valve into

a discharge system. Where more than one pressure relieving device discharges into a common system, built-up back pressure resulting from the operation of one device will act as a superimposed back pressure on the other devices.

Accumulation The term refers to the vessel to be protected and not to the safety valve. It is the pressure increase over the design pressure of the vessel when the valves discharge at rated capacity. Accumulation is the same as over-pressure when the valve is set at the design pressure of the vessel.

Over-pressure The pressure increase above set pressure, at which the discharge capacity of a safety valve is attained. The term refers to the safety valve and not to the vessel to be protected.

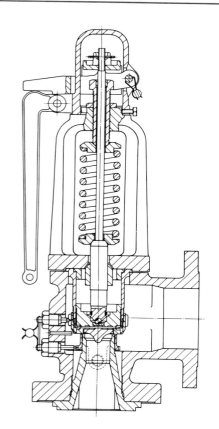

Fig. 5. Open bonnet safety valve

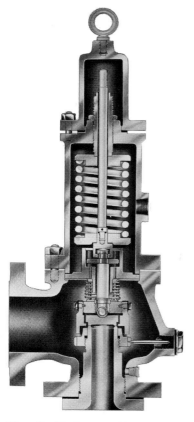

Fig. 6. Safety relief valve with balancing bellows

AUTOMATIC PROCESS CONTROL VALVES

The self acting regulator or reducing valve is generally acknowledged to be the forerunner of the modern automatic process control valve. Sensing from the line fluid and utilizing the pressure energy available, the valve controlled flow to give a regulated condition either upstream or downstream.

The valve controls flow by absorbing pressure from the line fluid, the quantity passing being a function of the pressure drop across the valve. Movement of the valve plug in relation to the seat provides an orifice of variable area that is utilized to regulate the flow from a minimum to the maximum. The prime function of a control valve is to regulate the quantity of line fluid and while tight closure may be obtained, this should be a secondary consideration in the choice of equipment.

Motive power for these valves comes from an actuator mounted directly on the body assembly. Early developments resulted in the pneumatic spring and diaphragm actuator which on attachment to the valve made remote operation possible. Controllers producing a pneumatic output enabled simple control loops to be established (Fig. 4), paving the way for the highly complex systems in use today. Pneumatic, hydraulic, and electrical means, or combinations of these, are currently used to provide the power for positioning the valve plug at the desired opening. Positioning accuracy and speed of movement are better than ever, with reliability of a very high order.

Irrespective of the medium used to power the actuator, two supplies are generally needed. The first is the control signal giving the command, while the other is the means to generate the movement required. In some cases the control signal may also provide the power for movement.

Frequently used in an automatic process control loop, the valve is the final control element. The sensing and detecting elements are located in another instrument from which the valve receives its command signal.

PROCESS CONTROL VALVE COMPONENTS

Automatic process control valves are produced in many sizes and configurations, all however having three basic components:

Body	Pressure containing component.
Trim	Actual control portion generally comprising a plug and a seat.
Actuator	Power source for positioning the plug against the fluid forces.

VALVE BODY

Since the body is the pressure containing portion of the valve, its design and construction have to be such that it will withstand the effects of pressure and temperature of the line fluids which it may be called upon to handle. The variety of solutions being handled in modern processes demands that care be taken in the selection of valves for particular applications, if satisfactory life in service is to be assured.

To cater for the differing needs of industry, a number of materials of construction are available. Most common are cast iron, bronze, carbon steel, and stainless steel. Aluminium, high nickel alloys, and proprietary brands are often available for special applications but it is wise to check with the manufacturer as to whether or not they can be supplied.

The majority of valve bodies are of cast construction. Forgings, too, are used and in the case of small valves bodies are frequently machined from barstock material. Choice of method depends on the manufacturer concerned but generally forged and barstock valves are only made in small sizes (up to 50 mm).

Body end connections, where the valve fits into the pipeline, can be provided in a variety of types. Most frequently supplied are those with flanged ends to ANSI, BS, DIN, or other standard specifications. Also available are screwed end with tapered female threads, weld end, and socket weld. Flangeless connection with through bolting between mating pipeline flanges, is increasing in popularity; long associated with butterfly valves, its advantages are being applied to other designs.

Many types of valves have been used as control valves with varying degrees of success but for many years the most commonly used type was the globe style with single or double seat plug, top and bottom guided. However, cage designs (Fig. 1) have increased in popularity to the point where they are now an industry standard. Other styles of sliding stem valves include split body globe (Fig. 2), angle, slant, or 'Y' body, and diaphragm, while those with semi-rotary stem movement include butterfly, ball, control ball, and plug. Variations of ball and disk shape and method of rotation have developed, like the example in Fig. 3, which has an eccentrically rotating

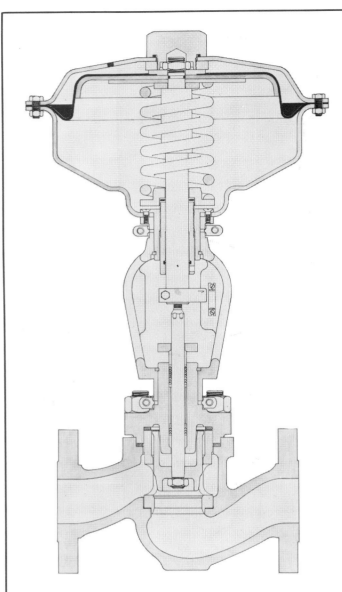

Fig. 1. Automatic process control valve, spring opposed, diaphragm actuated and cage guided

Fig. 2. Split body control valve with piston actuator

Fig. 3. Rotary type control valve featuring spherical section plug attached to operating shaft by flexible arms

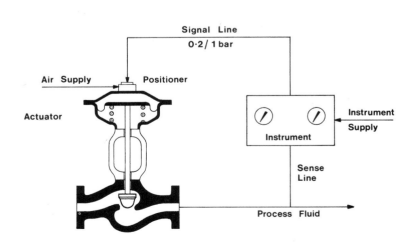

Fig. 4. Simple control loop

spherical plug, attached by flexible arms to the hub on the rotating shaft and contained within a free flow flangeless body. It has been found that there are situations where each will perform better than the others but globe valves are still the most frequently supplied types. Valves are frequently designed with a bonnet as a top closure. This member has guiding for the plug, with the upper portion containing a stuffing box through which the plug stem enters. In some styles, where the plug is guided at the top and bottom, a closure at the bottom end is the blind flange having the guiding for the lower end of the plug. In cage trim types blind flanges are not required, thus enabling a smoother flow passage to be obtained. It is usual practice to make these components of the same material as the body. Accuracy of alignment between the moving and fixed parts is essential: consequently, spigot location between body parts is normal. Sealing gaskets maintain pressure tightness, the type depending on the conditions of service. Low alloy steel or stainless steel studs and nuts secure the assemblies.

The bonnet or upper part of the valve body generally contains the means for mounting a yoke with actuator or an actuator direct. Alignment is assured by spigot mounting. Where the line fluid has a high temperature an extension to the bonnet is provided. Both plain and finned extensions are used, serving to keep the packing clear of the temperature affected zone and giving a longer service life.

It is the function of the gland packing to provide the maximum of sealing with the minimum of friction. Standard packings are in a number of materials, including braided asbestos impregnated with PTFE and machined PTFE chevron packing rings. Sometimes a spring is used beneath the chevrons to energize the assembly. Grease lubricators can be supplied and are a useful addition where the packing is not long term self lubricating. They should not be used with pure PTFE or pure graphite packing. Where lubricators are fitted it is essential that they are worked regularly in order to prevent the lubricant hardening and damaging the stem. Careful packing and adjustment will minimize stem friction, which is particularly relevant when the valve is fitted with a spring and diaphragm actuator. Piston actuators are less sensitive to this problem.

In the case of fluids which are dangerous, toxic, or costly, special sealing arrangements are available. One design has a deeper than normal stuffing box into which two independent sets of packing are introduced. The void space between is either pressurized by an inert gas or vented to a safe area. Where no leakage at all is permitted a bellows seal is needed; this takes the form of a metallic convoluted bellows, one end being welded to the valve stem and the other to the body assembly. The seal obtained by this method is absolute. A normal gland is provided as a backup in case of bellows failure.

VALVE TRIM

This usually means those internal parts controlling the flow and in physical contact with the line fluid, i.e., plug, seat(s), bushes, guides, cages, and stem.

The amount of fluid passing is regulated by the movement of the plug relative to the seat, which may be linear or rotary depending on the type of valve being considered. A powered actuator provides the force necessary to move the plug and maintain it in the desired position against the line forces.

The flow characteristic of a valve, i.e., variation in flow relative to plug movement, is of three basic forms, equal percentage, linear, and semi-throttle (see Fig. 5). This is achieved either by shaping the plug itself, having shaped holes in cages, making the seat to a particular profile, or by external means such as a cam in the valve positioner.

Plugs are machined to close tolerances with concentricity of a high order. Accuracy of alignment is assured, assisting good shut-off when closed and interchangeability when replacements are needed.

Seats are located in the body assembly, either screwed or clamped in position and generally spigot-located for accuracy. Sealing gaskets may or may not be fitted, depending on the manufacturer concerned. For arduous service, seats may be welded into the body to prevent slackening. To combat leakage behind the seat, a light seal weld between seat and body can often be provided.

Materials of construction for control valve trim are very important. Most frequently used is 18−8−3 austenitic stainless steel. A well tried and reliable alloy, it has good strength with an excellent corrosion resistance to many industrial liquors. Other materials used are the hardenable grades of stainless steel, i.e., ferritic, martensitic, and ferritic-austenitic, the latter being the most highly corrosion resistant. All of these grades have good strength in high temperature and low temperature (sub-zero) conditions, with corrosion resistance to many chemical solutions. Where problems of erosion are likely to be encountered the use of cobalt alloys, having a high degree of hardness, is recommended. Two forms are available. Plugs and seats can be made from the solid material: while expensive, this method provides the maximum resistance to erosion of the trim. Alternatively, a localized application of these alloys on to the stainless steel base material at the critical zones can be provided. Welding and spray−fuse techniques ensure good adhesion with protection provided only where it is needed. Guide bushings, and cages too, are available in hard materials. Where dirty line fluids or high temperatures are encountered this is often the only way in which a satisfactory service

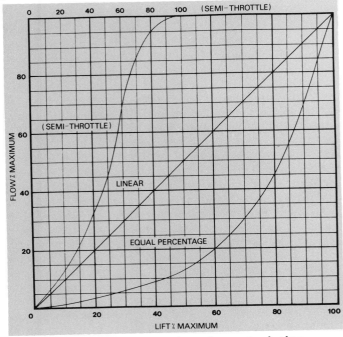

Fig. 5. Inherent valve characteristics

life can be obtained. Although the use of these materials is expensive in terms of first cost their use can be justified by the ability to maintain control over longer periods, thus helping to avoid costly unscheduled downtime. Tight shut-off should not necessarily be expected of a control valve. Indeed, it is possible to lap the seats in a double seat valve so that the leak rate at ambient temperature is virtually nil. At high temperatures the effects of differential expansion make tight closure unattainable. Changes of temperature, either high or low, can have a considerable effect on the leak rate.

In a single seat valve the problem is not so acute, the temperature change affecting mainly the distance between the plug and seat. This is absorbed by a slightly greater or lesser movement of the actuator, with the plug still being firmly seated.

Leakage rates of control valves are held to agreed limits by the manufacturers. For double seat valves the maximum leak rate is no greater than $0 \cdot 5$ per cent of the maximum rated C_v of the valve, while for single seat valves it is no greater than $0 \cdot 01$ per cent. Leak testing is generally done with compressed air at $3 \cdot 5$ bar with atmospheric pressure downstream. Testing is performed at the ambient workshop temperature.

Where tight closure is necessary, many manufacturers offer a resilient seating in synthetic material, to give bubble-tight or drop-tight shut-off. Depending on the design, the sealing member may be fitted to the plug or the seat. As the plug and seat close, the seal deforms, preventing any fluid passing. Normal materials for duty in this respect are PTFE and the better grades of synthetic rubber. In considering the use of synthetics it should be remembered that the pressure, and particularly the temperature, of the line fluid can be critical. The nature of the fluid can also have an important bearing on the suitability and life of a particular resilient material. Valve manufacturers have wide experience with problems of this nature and are always willing to offer advice or assistance in cases of difficulty.

CONTROL VALVE SIZING

To get the best from a control valve it is important that it is correctly sized, having regard to the conditions of the flowing media. While the equation for incompressible fluids is generally acceptable throughout the industry, there are variations in the formulae for gases and vapours. Reference should therefore be made to the valve manufacturer concerned for confirmation on the calculation methods employed. Reference may also be made to British Standards to assist in finding common ground so that the user will be able to make a better comparison between competing products.

International standards are currently being worked on under IEC control.

CONTROL VALVE NOISE

The control valve as a noise source is becoming more important with the increasing awareness of noise pollution and its effects on health. Noise in a valve originates in one of three ways: vibration of the valve internals, cavitation in liquid flow, and aerodynamic sound in gaseous flow. Of these, the last gives most cause for concern. Calculation methods for estimating cavitation and aerodynamic sound are available from manufacturers but universal standards are not available at time of publication.

VALVE ACTUATOR

Many types of actuator are available to power valves, enabling them to become part of a control system however simple or complex it may be. Pneumatic, hydraulic, and electric, or combinations of any two of these, form the majority of those in service.

Most frequently used is the pneumatic version. Proved over many years, this type has advantages of good power output, simplicity, reliability, and the use of a clean, safe operating medium. In petroleum refining and other processing industries where fire and explosion hazards are present, air powered equipment is widely used.

Pneumatic actuators are of the spring and diaphragm or piston operated types. Both are versatile pieces of equipment with features proved in service, in particular, the variety of fail safe modes which can be supplied.

With the spring and diaphragm design, air pressure introduced into the head acts on a flexible diaphragm. Movement of the diaphragm compresses a rate spring to which is connected the actuator spindle. A change of rate spring will give a different stroke for the same air signal range or allow another air signal range to be used for the same stroke length. Manipulation of the standard spring ranges available makes it possible to obtain a high initial thrust or high final thrust in the cases where this is needed, i.e., to close off the valve under conditions of minimum air signal and maximum air signal, respectively.

Springless actuators, in which the rate spring is replaced by an air pressure, are available but this design has never achieved great popularity.

Piston actuators, as the name implies, have a piston moving in a cylinder with air pressure each side of the piston. With these designs it is necessary to fit a positioner to ensure accuracy of output position unless they are used for on−off applications. Features of these units are high output force, great stiffness, and compact size.

Pneumatic control instruments generate air signal pressures of $0 \cdot 2 - 1$ bar; others such as $0 \cdot 4 - 2$ bar may also be encountered, as well as split range values $0 \cdot 2 - 0 \cdot 6$ bar and $0 \cdot 6 - 1$ bar. On receipt of an increasing signal pressure the actuator stem may extend (direct action) or retract (reverse action), depending on the valve type employed and what is required of it. A spring and diaphram actuator can operate direct from an instrument output but its speed of operation is slow. Piston actuators generally require a positioner when being controlled by an instrument. To get the best out of any actuator it is desirable to fit a positioner; not only is speed and accuracy improved but maximum power is developed throughout the stroke.

Fail-safe action is desirable in many situations and three modes are possible, i.e., on failure of the operating air supply the actuator stem can be made to extend, retract, or lock. In terms of a valve action this means air–fail–close, air–fail–open, or air–fail–lock. Which is chosen depends on the process or operational requirements; the user must decide which is best suited to his system as a whole.

A spring and diaphragm actuator has a built in fail-safe device in its spring. Piston types may have a helper spring for this purpose or, alternatively, internal pneumatic means or an external volume tank to achieve the desired movement on supply failure.

Hydraulic actuators are compact and powerful, often being the only way to overcome high stem loadings. Utilizing double acting cylinders, they have great stiffness with smooth motion. Since very few plant installations have an hydraulic main, a power pack must be supplied to provide the motive power; consequently, most applications are electro-hydraulic. Fail-safe action involving valve movement usually requires an hydraulic accumulator or an emergency electrical supply, otherwise the valve will remain at the last set position before power failure.

Electrical actuators are frequently used to power valves in remote situations where only an electrical supply is available. A great advantage being the ability to transmit signals over long distances with minimum trans-

mission losses. Although expensive, they are immune from freeze-up in very cold situations. Modern flameproof housings and weatherproof enclosures enable them to be installed in hazardous locations in an outdoor or exposed environment. Computer control of processing with all electric plants becomes increasingly attractive but fail-safe action has to be considered; as with hydraulic actuators, an emergency power supply is necessary if valve stem movement is needed when the power fails.

Hand operated override mechanisms can be furnished with most actuators irrespective of the power media. These provide an additional safety feature for plant personnel enabling flows to be maintained while difficulties are overcome. Both top and side mounted styles are seen, the type depending on the manufacturer's preference. Each perform the same function of allowing valve stem position to be under manual control should the need arise.

For accuracy of control, a positioner should always be fitted. With a built-in feedback mechanism the valve stem is always in the correct position relative to the input signal. Stem position accuracy is in the order of $0 \cdot 1$ per cent of the total movement. Any variation in position due to fluctuation in line forces brings a restoring force into play, returning the stem to the correct lift.

THERMOSTATIC VALVES AND MIXING VALVES

The term 'thermostatic valve' is not clearly defined but generally refers to a valve which is controlled automatically by some form of heat sensitive device.

In one form of temperature regulating valve the method of operation is very similar to that of pressure control valves inasmuch as the main valve is actuated directly or indirectly by a signal provided by a detector. In the temperature regulating valve a heat sensitive device, which may be of the solid expansion type or the liquid filled bulb type using non-volatile or volatile liquids, is employed to control the flow through the main valve and thereby regulate the temperature of the flowing medium.

Mixing or blending valves may also use a thermostat element to control the relationship between the cold flow and the hot flow, so as to provide the required outflow temperature.

Mixing valves may also be of the manually controlled type. A single handle controls both the hot and cold supplies simultaneously and allows the required outlet temperature to be selected.

FLOAT OPERATED VALVES

These valves are used principally on the inlets to service reservoirs or tanks for the purpose of maintaining a pre-determined water level. The most commonly used version consists of a plain spherical float attached by some form of lever mechanism to the valve element. The float follows the water level in a partially submerged state, causing the valve to open and admit more water with falling water level and to close as the water rises. The reverse action – valve opens as float rises and closes as float falls – can also be provided for the discharge of water accumulating in tanks.

Both angle and in-line types of valves are used in a variety of designs. Two examples of conventional types of valves are shown in Figs 1 and 2.

A design of valve particularly well suited for higher pressure applications is illustrated at Fig. 3. The streamline flow pattern provides smooth handling of high velocities without vibration, erosion, or noise. The float action has generally only to operate a pilot valve, incorporated in the design, giving sensitive response to small changes in the water level and drop-tight closure.

FLOAT ARRANGEMENT

The simplest arrangement consists of a spherical float attached to the valve actuating lever and operated directly by the changing water level in the main tank or reservoir.

Possible oscillatory action of the valve due to surface turbulence in the reservoir can be avoided by the use of a separate float tank, which may take either of two forms:

Plain Stilling Tank
A typical arrangement is shown in Fig. 4. A cheese float mounted on a central tube and connected to the valve operating lever slides vertically on a guide rod secured to the base of a tank. The tank base is perforated for communication with the main reservoir.

Siphon Type Stilling Tank
Figure 5 shows a typical arrangement. The tank fills through a siphon pipe and empties through a subsidiary float operated valve. It cannot begin to fill until the water in the main reservoir reaches top level, when it fills completely and the valve moves from the full open to the full closed position in a continuous stroke. Similarly, the opening movement cannot begin until low water level is reached in the reservoir, when the tank empties and the full opening stroke is effected.

In addition to ensuring still water for the float, this device also eliminates the unfavourable condition of prolonged operation in part open positions.

44

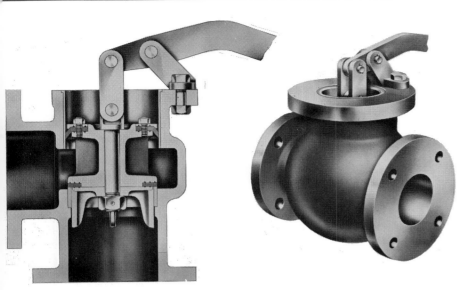

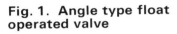

Fig. 1. Angle type float operated valve

Fig. 2. Globe type float operated valve

Fig. 3. Streamline design of float operated valve for higher pressure applications

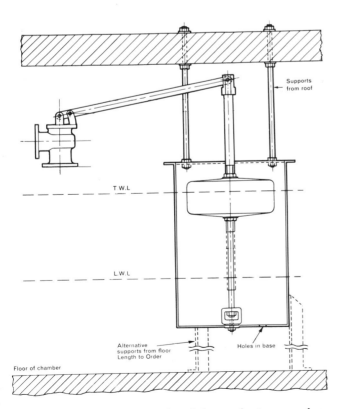

Fig. 4. Plain stilling tank with angle type valve

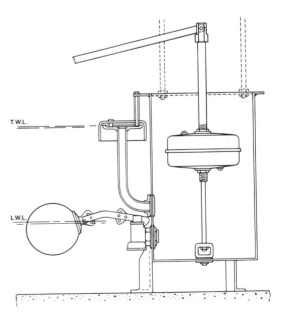

Fig. 5. Siphon type stilling tank

AIR VALVES

Air valves are regularly used to provide for the following requirements on water mains:

(1) To release air when the main is being filled, and to close and remain closed when the pipe is full to prevent loss of water. Also, to open and admit air when the main is being emptied.

(2) To release air accumulating under pressure during normal working conditions in the pipe, again without loss of water.

Conventionally, this operation is effected automatically by means of a ball float working in conjunction with an orifice of appropriate type for the duty.

Single Air Valve with Large Orifice (Fig. 1)
For requirement 1. With the main empty the ball float is at the bottom of its travel and the orifice is open. As filling proceeds, air is discharged until the rising water level floats the ball on to the orifice seating where it is subsequently held by line pressure, thus sealing the outlet. On emptying the main, when the pressure falls to near atmospheric the ball drops and allows air to be admitted through the orifice.

Single Air Valve with Small Orifice (Fig. 2)
For requirement 2. Under operational conditions the ball is normally held against the seating of the small orifice.

As air accumulates in the valve chamber the water level is depressed, until loss of buoyancy brings the ball from its seating. Air is then discharged and the consequent rise in water level brings the ball up to reseal the outlet.

To increase the permissible working pressures of small orifice valves the ball float may be made to control the orifice through a lever mechanism, thus multiplying the operating force. For the higher pressures a counterweight device is added, to compensate for the necessary heavier construction of the ball.

Double Orifice Air Valve (Fig. 3)
This valve has one large orifice for release and admission of air when filling and emptying the main respectively, and one small orifice for release of air accumulating under normal working conditions.

Triple Orifice Air Valve
Basically a double-air valve for large and small orifice duties but with an additional large outlet to provide extra capacity.

Isolating Valve
Provided, when required, to isolate the air valve from the main. May be separate or incorporated in the air valve.

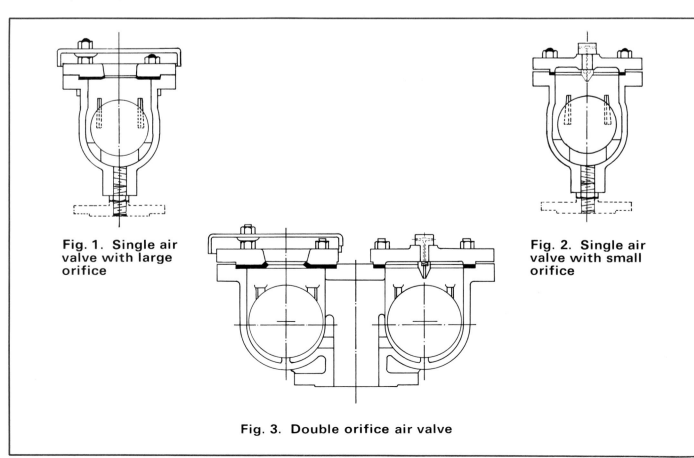

Fig. 1. Single air valve with large orifice

Fig. 2. Single air valve with small orifice

Fig. 3. Double orifice air valve

BELLOWS SEALED VALVES

Today there are many services in industry and elsewhere where nil permissible leakage to atmosphere is an absolute requirement to meet essential safety and operational demands. Prime examples are operations in the field of nuclear power, in the petroleum and chemical processing industries, in explosive or toxic services, and many other applications involving hazardous media. The absolute integrity of sealing demanded by such services is beyond the capability of conventional forms of glands and seals and this has led to the development of the modern bellows sealed types of valves.

As may be seen from Figs 1 and 2, the bellows unit is a flexible metallic membrane the bottom end of which is welded around the lower end of the stem and the top end welded to some part of the valve casing. A stem anti-rotation device is incorporated in the design to prevent torque being applied to the bellows as the valve is opened and closed. On some designs a shroud is arranged around the bellows to protect it as much as possible from harm by erosion.

The development of bellows has advanced greatly during the last decade or so and a life of up to a minimum of ten thousand cycles in operation is now designed for and expected.

As an assurance in the event of a failure of the bellows, for example by fatigue failure, it is not unusual when dealing with hazardous media to fit an additional back-up gland having conventional type packing or seals.

The principal metal used for bellows is stainless steel but other materials are available including bronze, 'Inconel', 'Nimonic', 'Monel', and titanium.

Theoretically, there is no reason why the bellows sealing feature should not be applied to most types of valves but at the present time the valves which are generally commercially available are the globe and gate types.

The bellows sealed valve can also play a considerable part in the conservation of energy, i.e., steam, fuel oil, etc., by virtue of its nil leakage to atmosphere. At the same time, of course, it considerably reduces the need and consequent cost of gland maintenance.

Fig. 1. Bellows sealed globe valve

Fig. 2. Bellows sealed gate valve

SOLENOID VALVES

A solenoid valve is a combination of two basic functional units.
  (1) A solenoid (electromagnet) with its plunger (core).
  (2) A valve containing an orifice, in which a disk is positioned, either to allow or to prevent flow of the medium being controlled.

The valve is opened or closed by movement of the magnet plunger (core), which is drawn into the field of the coil when the coil is energized.

TYPES OF SOLENOID VALVE

In the main, solenoid valves are covered by the following types of operation.

Direct Acting Valve

In a direct acting valve (**Fig. 1a and 1b**) the solenoid plunger (core) is mechanically connected to the valve disk and directly opens or closes the orifice, depending upon whether the solenoid is energized or de-energized. Operation is independent of line pressure or rate of flow and the valve will operate from zero pressure to its maximum rated pressure.

Internal Pilot Operated Valve

This type of valve (**Fig. 2a and 2b**) incorporates a pilot and bleed orifice and utilizes the line pressure for operation. When the solenoid is energized the pilot orifice is opened and pressure is released from the top of the diaphragm to the outlet side of the valve. This results in an unbalanced condition which causes the line pressure to lift the diaphragm and open the main orifice. When the solenoid is de-energized the pilot orifice is closed and full line pressure is applied to the top of the diaphragm via the bleed orifice, thereby providing a seating force to close the main orifice.

External Pilot Operated Valve

This valve (**Fig. 3**) is usually of diaphragm or piston operated construction, equipped with a three-way solenoid pilot valve. The pilot valve alternatively applies pressure to or exhausts pressure from the diaphragm or piston for operation of the main valve. Line pressure or a separate source of pressure can be utilized to operate the pilot valve.

The simple two-way valves briefly described above form the majority of solenoid valves used in industry. Other more complex types in use include three-way and four-way valves, manually and electrically reset valves, and a wide variety of special valves.

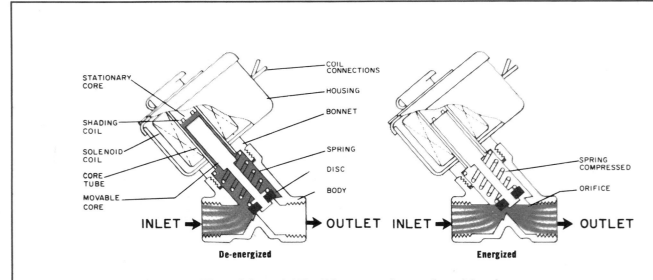

Figs. 1A and 1B. Direct acting solenoid valve

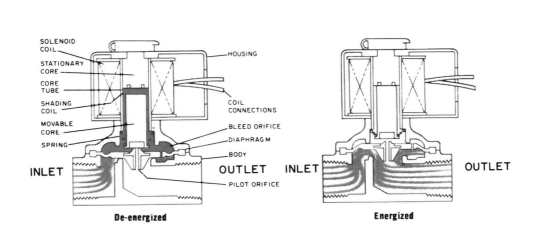

Figs. 2A and 2B. Internal pilot operated solenoid valve

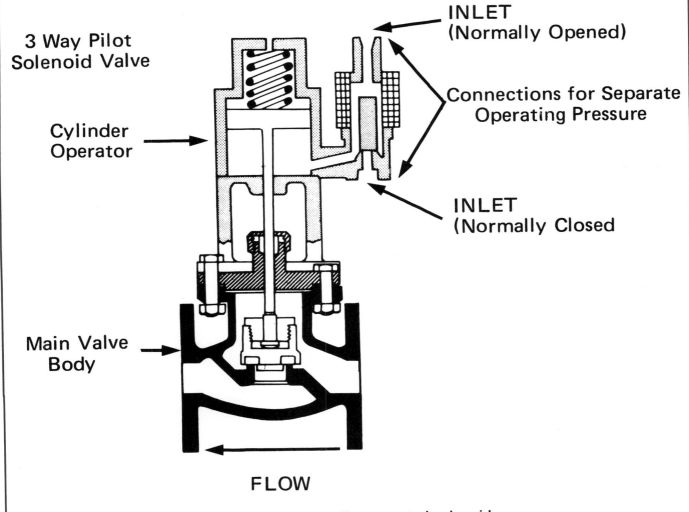

Fig. 3. External pilot operated solenoid valve

CRYOGENIC VALVES

Cryogenic valves are generally defined as valves operating at temperatures below −40 °C and are usually found to be handling liquefied gases such as nitrogen, oxygen, and natural gases at temperatures down to −196 °C.

Design Features

For such conditions, special design features are necessary and materials used require careful selection. Valves for service down to −40 °C are in an intermediate category and, provided the materials are suitable, standard designs often prove adequate.

The types of valves most usually employed for cryogenic services are gate, globe, and ball valves, although others can be used.

The distinguishing feature of most cryogenic valves is usually a gas column and extended bonnet arrangement between the main body of the valve and the operating handwheel or lever (see Fig. 1). The purpose of this is to allow a reasonable temperature gradient from the very low temperature of the valve up to the point of operation and this gas column will, at some point and beyond, be at a temperature at which the liquefied gas ceases to be liquid and reverts to the gaseous state.

It is therefore important in the design of the valve that this column is at all times vented (usually to one side or other of the valve) so that dangerously high pressures are not generated if the temperature rises in the column.

In the design and manufacture of cryogenic valves very careful attention has to be paid to the differential expansion and contraction with temperature change of the various components, which will generally be manufactured and inspected at ambient temperatures.

Materials

Materials employed must be suitable for the very low temperatures, ordinary steels being quite unsuitable since they suffer embrittlement with reducing temperatures. The most commonly used metals are austenitic stainless steels, bronzes, and cupro-nickels, none of which exhibit this effect. In fact, the bronzes and cupro-nickels are the more stable materials as temperatures are reduced.

PTFE and similar non-metallic materials may be used for facings of seating elements such as gaskets and disk facings, or inserts, and for gland packings.

Testing

Testing of cryogenic valves can only truly be carried out at the very low temperatures at which they are intended to operate. Since such tests are usually conducted using liquid nitrogen as the cooling agent and with helium as the test fluid, they are not inexpensive. It is therefore quite common to carry out type testing on a limited number of valves in a batch and merely to conduct normal ambient temperature tests on the rest, alongside strict quality control of materials and dimensions, which ensures that the tested valves are fully representative of the batch.

Fig. 1. Cryogenic gate valve showing clearly the gas column between the bonnet and the gland

PINCH VALVES

Basically, the pinch valve comprises a reinforced sleeve of natural rubber or some synthetic elastomer that is pinched or flattened to produce a closure. The sleeve may be exposed with metal ends for coupling to piping or it may be completely encased in metal.

Valves may be manually operated or power operated. Figure 1 shows a hand operated valve employing a simple wheel and screw arrangement to pinch the sleeve and produce closure. Power operation is often pneumatic or hydraulic but electrically operated devices are also used. An example of a pinch valve with opposed pneumatic diaphragm actuators is shown in Fig. 2.

In one concept the sleeve is encased and pressurized liquid or gas is introduced between the metal casing and

sleeve so that the sleeve walls are squeezed together, thus shutting off the flow. Figure 3 illustrates a valve of this kind in which compressed air is used to pinch the sleeve.

Although the elastomer materials of which the sleeves are made restricts the operating temperatures and pressures of pinch valves they have a wide variety of applications. Because the operating mechanism is completely isolated from the service fluid and there is unobstructed full bore flow, pinch valves are well suited for use with slurries, liquids containing large amounts of suspended matter, and solid materials in powder form. They can cope with corrosive services and, with sleeves made from the appropriate materials, are suitable for use with edible and medical products.

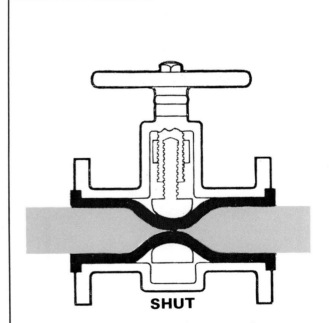

Fig. 1. Handwheel operated pinch valve

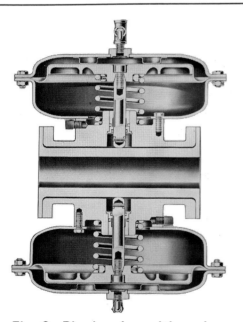

Fig. 2. Pinch valve with spring opposed pneumatic diaphragm actuators

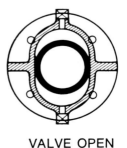

VALVE OPEN

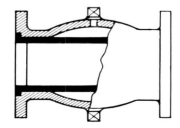

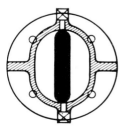

VALVE CLOSED

Fig. 3. Pinch valve for operation by compressed air

GAS DUCT ISOLATORS

These valves are intended for shut-off duty in process plant gas ducts, the sizes of which may be up to 100 square metres. Various designs of gate, (including guillotine), flap, and louvre type valves are available, the selection depending on such parameters as pressure differential, temperature, gas analysis, dust burden, and space limitations. Gas tightnesses up to and including 100 per cent are possible.

Temperature and gas analysis determine the choice of materials, while the dust burden and nature of the dust are critical factors in the choice of valve type. Heavy dusty burden may prevent full closure of flap or louvre types and necessitate the use of gate or guillotine valves. Even then it may be necessary to provide special provision for dust clearance to achieve reliable operation.

Gas duct isolating valves are generally custom built to suit specific operating requirements and are commonly of fabricated construction. Figure 1 illustrates a typical guillotine type of unit as installed in a power station for electrostatic precipitator isolation.

Regular applications of isolators provide a safe and economic means of pollution control to atmosphere, for example in power stations, steel works, and cement plants, and the isolation of primary and ancillary plant to allow for on-load inspection and maintenance.

Isolators are also an increasing requirement in total energy schemes where the application of suitable insulated designs of tight sealing isolators can effectively minimize heat losses by radiation and convection.

Fig. 1. Duplex guillotine isolator incorporating peripheral air barrier

GENERAL FEATURES

END CONNECTIONS

The choice of a type of end for connecting a valve to its associated pipework is dependent upon various factors. Some typical examples are the nature, pressure, and temperature of working fluid, and the frequency of dismantling the pipeline itself or of removing the valve from the line. Additionally, the end connections are to some extent governed by the type of pipe normally employed for a particular duty. These may range from lead pipe for domestic water supplies to solid drawn creep resistant steel pipes for conveying fluids at high pressures and temperatures. Some of the more commonly used types of connections are as follows.

TYPES

Threaded

Male threads of various forms may be used for special purposes, but as a rule threaded end valves have female pipe threads, either tapered for assembly to taper threaded pipe or parallel for assembly to taper or parallel threaded pipe. In the case of taper to taper and taper to parallel connections, the pressure-tight joint is made on the threads. For parallel to parallel connections, the pressure tight joint is made by compressing a grummet or gasket against the end face of a valve. Threaded ends, usually confined to pipe sizes 150 mm and smaller, are widely used for bronze valves and to a lesser extent for iron and steel valves. They are not recommended for severely corrosive conditions.

Flanged

Although flanged ends are used on quite small valves they are found most generally on the larger sizes of valves and used for practically all pressure–temperature ratings. The bolted form of connection between valve and piping facilitates installation of the valve in the line and is convenient should it be necessary to remove the valve for inspection or maintenance. When fitted on bronze valves the flanges usually have plain (flat) facings. Iron valves may have plain or raised facings and steel valves may have plain, raised, female, tongue, groove, or ring joint types of flanges.

Gaskets

To ensure a tight seal it is usual to fit gaskets between the end flanges of the valve and piping. The three most regularly used types are: (1) flat ring gasket with a wide face extending to the inside of the flange bolt holes or a narrow face conforming to the dimensional requirements of a male–female type of flange joint; (2) full face gasket which extends across the whole face of the flange, the bolts passing through holes in the gasket; (3) solid ring gasket in which a metal ring, usually of oval or octagonal section, fits into machined grooves in the mating flange faces. Other forms of flange sealing employed include PTFE cord and gasket tape, and sealants applied in a liquid form.

Most gaskets are manufactured from one or more of the following materials: natural and synthetic rubber, cork, asbestors, paper, cotton, linen or synthetic woven fabric, plastics (chiefly PTFE), and metals.

In the more rigorous applications it may be desirable to protect the basic gasket material or to provide additional strength. This can often be done by the use of a suitable thin metallic envelope surrounding the asbestos compound. In other cases added protection (but not strength) can be provided by the use of PTFE as an envelope for the basic material. This is particularly suited to severely corrosive conditions.

Finally there are the rather more recent spiral wound gaskets which can be selected for both strength and resistance to attack by the fluid. These consist of a spirally wound metallic strip with an interspersed filler of either asbestos or PTFE. By the use of stainless steel or other alloy for the service, and the selection of the filler, a very strong and durable gasket is provided for most conditions.

Socket Weld

The ends of the valve are socketed to receive plain-end pipes and the joints secured by circumferential welds. Socket weld ends are used usually on the smaller sizes of valves for higher pressure–temperature applications in pipelines not requiring frequent dismantling and on severely corrosive services for which threaded ends would be unsuitable.

Butt-weld

In this case the ends of the valve are bevelled to match machined bevels at the ends of the mating pipes and a circumferential weld is made at the abutted mating bevels. 'Backing rings', which are basically sleeves fitting inside the pipe, are sometimes used to align the pipe and valve bores. They also prevent 'icicles' and weld spatter from entering the pipeline. Butt-weld ends are generally used only for the more severe applications where frequent dismantling is not required.

Compression

This type of valve end has a socket to receive the pipe and is fitted with a screwed union nut. The joint is made by the compression of a ring or sleeve on to the outside of a plain-end pipe or by compressing a preformed portion of the pipe end. As a rule compression ends are used with copper tubing but they are also applicable to the smaller sizes of steel pipes.

Capillary

Here the valves are soldered to the mating pipe. The ends of the valve are socketed and machined to close tolerances to receive plain-end pipes and the joints made by the flow of solder by capillarity along the annular spaces between the sockets and the outsides of the pipes. Capillary ends are commonly used with copper tubing but their use is limited owing to the comparatively low melting point of the solder.

Socket

The ends of the valve are socketed to receive the plain ends of the piping and the seals made by the insertion of yarn ring jointing, caulked with lead. Other types of socket ends use rubber sealing rings with some form of gland or locking ring. Socket ends are normally associated with cast iron valves for water services.

Spigot

The type of socket used in the coupling or on the pipe end determines the form of the spigot ends. For cast iron pipes with lead joints the spigot end is provided with a raised band; for screwed and bolted glands and other forms of mechanical joint the spigot end is prepared to suit the joint. For asbestos cement connections the spigot end is finished plain in the same way as the pipe. Spigot ends are normally associated with cast iron valves for water services.

BODY BONNET JOINTS

Note. In this section both bonnets and body covers are included within the meaning of the term bonnet.

Primarily the bonnet provides a closure for the valve body. It has to withstand the pressure and temperature conditions of the service fluid and usually has to be removed to gain access to the working parts. The design of the body–bonnet joint should therefore provide not only a pressure sustaining leakproof seal but also allow for ease of dismantling.

TYPES

The three most common types of connections are threaded, threaded union ring, and bolted flange, but other designs are available, providing the user with a range of joint characteristics.

Threaded

This is the simplest form of construction (Fig. 1), the bonnet being screwed directly into the body. It is often used on smaller sizes of valves for moderate pressures and where frequent dismantling is not required. Depending upon the detail design, the joint may be sealed by metal to metal contact or by the use of a gasket.

Threaded Union Ring

The bonnet is connected to the body by means of a threaded union ring in a manner similar to the conventional pipe union (Fig. 2). The separate union ring applies a direct load on the bonnet to make a strong and reinforced joint between bonnet and body.

The turning motion used to tighten the ring is spent between the shoulders of the ring and the lip on the bonnet, so the area of sealing contact between bonnet and body is less subject to wear from frequent opening of the joint. In some cases a gasket may be fitted to seal the joint.

The union ring type of connection is very convenient where valves require frequent inspection or cleaning. Dismantling is easily done with less risk of injury to the joint faces.

Although suitable for quite high pressure services, the union ring type of connection is impractical for large valves and its use is therefore usually restricted to valves 100 mm in diameter and smaller.

Bolted

Generally used on the larger sizes of valves and on services where corrosive fluids and high temperatures and pressures are involved (Fig. 3). Adaptable to all kinds of gaskets and ring type joints. The multiple bolting provides equalized gasket pressure and maintains a tight bonnet joint seal.

Pressure-Seal Fig. 4

In this design the line fluid pressure is utilized to seal the joint. The actual joint is inside the valve and is sealed with a wedge-shaped seal ring. Internal fluid pressure acting upon the entire underside of the bonnet forces it against a seal ring which is then wedged between the bonnet and the body, forming a pressure-tight metal-to-metal joint. The design ensures that the sealing pressure is always many times greater than the fluid pressure and the higher the internal pressure the greater is the sealing pressure.

This joint is particularly suitable for high pressure and temperature services.

Seal Welded Fig. 5

Body and bonnet are joined together in the style of the threaded connection or bolted connection and the periphery of the joint is then seal welded.

The threads or bolting carry the entire mechanical load, the seal weld simply guarding against any leakage. The weld can be easily ground off if it is required to dismantle the valve for maintenance purposes.

Breech Lock

The bonnet is secured to the body by interlocking breech lugs which carry all the thrust imposed on the bonnet by the pressure of the service fluid. To assemble the valve the bonnet is lowered into the body and then rotated 45° to engage the body and bonnet lugs. The joint is completed by making a small seal weld between the body and a flexible steel ring on the bonnet. This weld can be readily chipped out if the valve has to be dismantled for maintenance purposes.

This joint is well suited for high pressure and high temperature services.

Fig. 1. Threaded joint

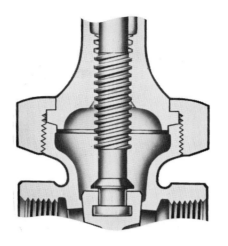

Fig. 2. Threaded union ring joint

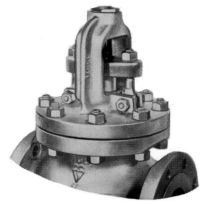

Fig. 3. Bolted joint

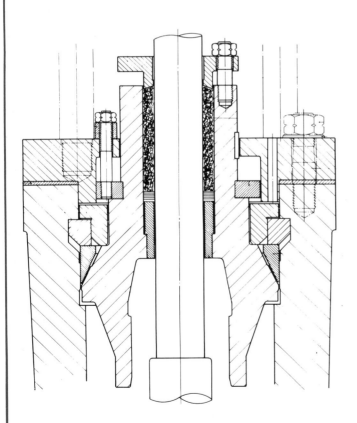

Fig. 4. Pressure-Seal joint

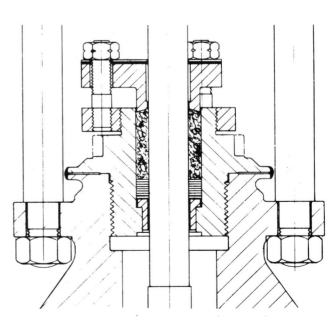

Fig. 5. Threaded and seal welded joint

STEM SEALING

An important consideration in valve design is the sealing of the valve stem. This may have a turning or sliding movement or a combination of both and the sealing must be capable of contending with the movement of the stem while ensuring tightness against the pressure of the fluid inside the valve. The temperature of the fluid may also be a factor influencing choice of sealing, and toxic and dangerous gases require special consideration.

TYPES

Compression Packings

The type of stem sealing found perfectly satisfactory for most liquid and gas services is the conventional stuffing box with soft (compression) packing, typified in Fig. 1. A spigoted gland, bolted or threaded, applies a compressive force to the packing and the resulting radial pressure of the packing on the stem provides the seal. To ensure fluid tightness the radial pressure must exceed the pressure of the system fluid.

Springs can also be used to create the compressive force on the packing and are useful where the gland assembly is inaccessible. They do not however provide the fine degree of adjustment needed for many critical duties.

Lantern Ring

On difficult applications it is not uncommon to incorporate a lantern ring into the gland area (Fig. 2). Where it is essential that there is no leakage of the service fluid to atmosphere a leakoff to a safe area can be arranged at the lantern ring or a barrier fluid can be introduced into the lantern ring area, usually at a pressure slightly in excess of the line condition. In hot climates, a lantern ring may be used as a cooling chamber or for relieving build-up of pressure inside the valve bonnet. It may also be used for the introduction of a further lubricant should this be desired.

Lantern rings can serve a number of very useful purposes but they may also provide a possible source of shaft scoring and should therefore be used only where the application makes their presence essential.

Automatic Packings

Packings in this category are sometimes preferred to compression packings for certain special services. There are two distinct divisions into which these seals may be classified, lip type and squeeze type. In both cases the sealing pressure on the valve stem is produced by the response of the seal to the system pressure.

Lip type sealing rings (also known as 'V' rings), illustrated in Fig. 3, have the same basic housing arrangement as for compression packing but do not necessarily require any externally applied force. Some adjustment may be provided by incorporating spacers between the gland flange and the top of the housing. Lip-tight rings are often made in PTFE for applications where chemical resistance is a major requirement.

The 'O' ring (Fig. 4) is the most popular form of squeeze type packing and can be housed within recesses of simple design and relatively small proportions. It is often used on small valves in place of the more traditional compression type of packing.

Packless Valves

For applications where there must be absolutely no leakage whatsoever to the outside, e.g., hazardous fluids, a number of valves that use a packless method of stem sealing are available.

One example is the diaphragm type, in which the bonnet assembly is sealed off from the fluid passageway by a flexible diaphragm clamped between the body and the bonnet. Another important type in this category is the bellows sealed valve, which employs a flexible metallic membrane to seal off the stem from the service fluid. Both diaphragm and bellows sealed valves are described more fully in the section 'Types of Valves'.

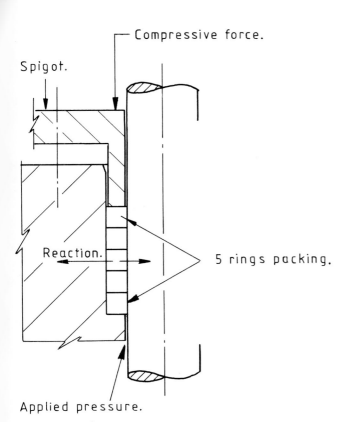

Fig. 1. Compression packing

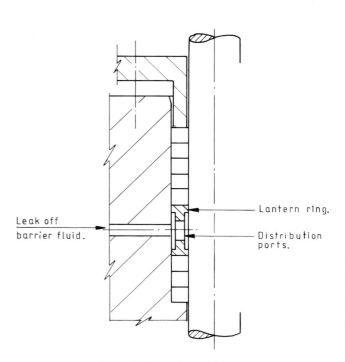

Fig. 2. Lantern ring

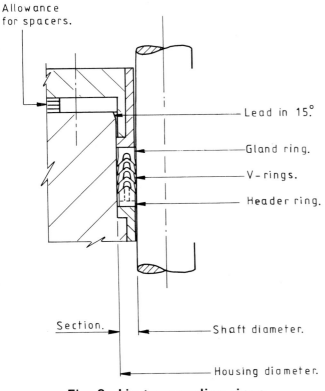

Fig. 3. Lip type sealing rings

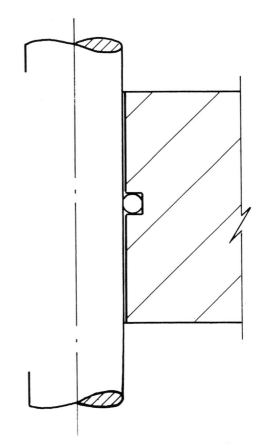

Fig. 4. 'O' Ring seal

MATERIALS OF CONSTRUCTION

A wide range of materials is available to meet the many different service conditions in which valves may be required to operate. It is useful to consider the question of material selection in relation to the function of the component, so that the material for the envelope of the valve (body and bonnet or cover) may be different from that of the trim (closing member, body seats, and stem). In this way, optimum economic life and performance can be obtained.

The materials most frequently used for valve bodies and bonnets are bronze, cast iron, and steel and some consideration of each of these is given in the following. Trim materials are considered separately.

It should be understood that these notes can be only of a generally indicative nature and in making a specific selection it is necessary to pay regard to manufacturers' specifications, national codes of practice, and local standards appropriate to the duty involved.

BODIES AND BONNETS

Bronze

The great majority of the vast number of valves supplied every year to industry are made from nonferrous materials, chiefly 'bronze' or brass. Bronze in valve manufacture is usually an alloy of copper, tin, lead, and zinc of the 85/5/5/5 or 87/7/3/3 type. If a zinc-free bronze is required this must be specified. The physical strength, structural stability, and corrosion resistance of bronze in particular makes it very suitable for a wide variety of applications in the mainstream of industrial requirements. Bronze also lends itself particularly well to economic large scale production of the smaller size of valves and, conveniently, the bulk of industrial applications is for valve sizes up to 100 mm.

Although bronze valves are used mostly for applications at relatively modest temperatures, some grades of bronze are suitable for temperatures up to around 280 °C. At the other end of the temperature scale the attribute of not becoming embrittled at very low temperatures, possessed by most copper alloys, has led to bronzes being widely used for cryogenic services such as liquid oxygen and nitrogen where temperatures below −180 °C prevail.

Cast Iron

Cast iron valves offer a considerable cost advantage and have a wide range of permissible service applications in practically every field of industry. They are commonly used on water, steam, oil, and gas services and find numerous applications in the handling of chemicals, dyestuffs, paints, textiles, and many other industrial products where a degree of iron contamination is of little or no consequence. The operating temperature range is usually from ambient up to around 220 °C.

Valves are also available in s.g. (spheroidal graphite) iron (also known as nodular or ductile iron), a form of cast iron in which the graphite is present substantially in nodular or spheroidal form instead of in flakes as in grey iron. This modification to the structure of the metal gives it mechanical properties superior to those of ordinary grey iron without detriment to its other characteristics. Consequently, valves made of s.g. iron can be used for operating pressures higher than those associated with grey iron valves.

Steel

Cast carbon steel valves were developed originally to cope with operating requirements beyond the capabilities of iron or bronze valves but the general serviceability and greater resistance to stresses, caused by thermal expansion, shock loads, and pipeline strains, of carbon steel valves, have resulted in their use being extended to services that are often within the scope of cast iron or bronze valves.

Bodies and bonnets of carbon steel or low alloy steels such as carbon–molybdenum and chromium–molybdenum are suitable for a great variety of services, including saturated and superheated steam, cold and hot oils, gases, and air. Operating temperatures up to around 540 °C are possible using carbon steel and over 600 °C for valves in low alloy steels. The latter steels also have greater physical strength than carbon steel at the higher temperatures.

Special carbon steels and low nickel alloy steels are used for subzero services not extending fully into the cryogenic zone and valves made of these materials are used on services such as brines, carbon dioxide, acetylene, propylene, and ethylene.

Stainless Steel

Stainless steels containing about 18 per cent chromium and 8 per cent nickel, 18–8 austenitic stainless steels, are regularly used as body and bonnet materials for services at elevated and subzero temperatures and for highly corrosive conditions. The addition of molybdenum to the basic type 18–8 steel and a slight increase in nickel materially increases its corrosion resistant properties and valves made of 18–10–3 Mo steel are used extensively in the chemical industry for handling acetic acid, nitric acid, alkalis, bleaching solutions, food products, fruit juices, sulphurous acid, tanning liquors, and many other industrial chemicals.

For use at elevated temperatures a further modification is made by the addition of niobium and this steel, known as 18–10–Nb, is suitable for temperatures up to 800 °C.

Austenitic stainless steels usually do not suffer embrittlement at extremely low temperatures, so valves in materials such as 18–8 and 18–10–3Mo are very suitable or operating on cryogenic services; instances are the handling of liquefied gases such as natural gas, methane, oxygen, and nitrogen.

Special Stainless Steels

Where conditions are too severe for the standard stainless steels the next group of interest is that of the more highly alloyed stainless steels. Probably the most common of these is '20' alloy, which contains about 29 per cent nickel and 20 per cent chromium with additions of molybdenum and copper. This alloy is extremely resistant to sulphuric acid over a wide range of concentrations and temperatures. In addition it will handle phosphoric and acetic acids under most conditions, especially where chlorides or other impurities are present.

Even more highly alloyed materials are available, such as 'Incoloy 825' and 'Carpenter 20Cb3', which may be required for the more extreme conditions.

There is also a growing use of duplex stainless steels (having a ferritic or austenitic structure) which contain 20 per cent or more of chromium and 5 per cent or so of nickel, with some molybdenum. These alloys are stronger and harder than the standard austenitic stainless steels and have better resistance to selective corrosion in the more severe conditions of sulphuric and phosphoric acids.

'Monel' Alloy

A high nickel–copper alloy with good general corrosion resistance. It is often used for valves handling alkalis, salt solutions, food products and many air-free acids, in particular sulphuric and hydrofluoric. 'Monel' alloy is extremely resistant to steam, sea water, and marine environments.

Nickel

Pure nickel offers even greater resistance to alkalis and salt solutions and is usually used where it is desired to ensure a very high degree of purity in the product handled.

'Hastelloy' Alloy 'B'

An alloy containing about 60 per cent nickel, 30 per cent molybdenum, and 5 per cent iron. It has exceptional resistance to highly corrosive mineral acids. All concentrations of hydrochloric acid can be handled at temperatures up to the boiling point, and for sulphuric acid 'Hastelloy' alloy 'B' is used extensively for the more severe conditions of concentration at temperatures above 70 °C. It can handle phosphoric acid under most conditions and is suitable for ammonium and zinc chlorides, aluminium, and ammonium sulphates.

In oxidizing atmospheres 'Hastelloy' alloy 'B' can be used up to approximately 800 °C, and at even higher temperatures in reducing atmospheres.

'Hastelloy' Alloy 'C'

The presence of 15 per cent chromium and 17 per cent molybdenum in this nickel based alloy renders it suitable for handling both oxidizing and reducing atmospheres up to 1100 °C. The resistance to hydrochloric, sulphuric, and phosphoric acids is good and the alloy will handle nitric acid under many conditions.

'Hastelloy' alloy 'C' is also highly resistant to wet chlorine and hypochlorites, sulphites, oxidizing salt solutions, and many other corrosives. Resistance to halogen acids such as hydrofluoric acid is also exceptional.

Plastics and Elastomers

Plastics and elastomers are used in valve construction where corrosion resistance and cleanliness are required and also for the quantity production of valves and components by moulding where the finished parts can be made accurately, with good appearance and surface finish.

The mechanical properties and range of temperature operation of these materials is less than that of most metals but with proper selection and design this is not a serious disadvantage for many applications.

Complete valves, usually of the ball, plug, or diaphragm type, are moulded in plastics, the following being the more common materials used:

polyvinyl chloride–unplasticized (PVC)
chlorinated polyvinyl chloride (CPVC)
polypropylene
polyvinylidene fluoride (PVDF)

These materials have excellent resistance to a wide range of chemicals and many compounds are approved for use on foodstuffs. Such valves are used in plastics pipework systems where connection can be made by solvent or fusion welding as well as with threaded or flanged joints.

The upper temperature limit of these materials varies from about 60 °C for PVC to 150 °C for PVDF. Actual operating temperatures and pressures are closely related and will vary with the valve size.

A variety of plastics is used for certain structural valve components, such as bonnets, handwheels and spindles. The following are the more common materials used:

polypropylene
acrylonitrile butadiene styrene
nylon
polycarbonate
polyphenylene oxide
acetal

Many of these materials possess good tensile strength and toughness and their properties may be enhanced by the addition of fillers, such as glass fibre. The use of plastics is generally restricted to relatively small valve components and excludes applications involving extremes of temperature.

Plastics are used for lining valve bodies where it is required to maintain the full working pressure of the valve and utilize the corrosion resistance and temperature properties of the plastics. Linings are usually moulded into the valve body and the body itself forms the outer part of a moulding tool.

Plastics linings are applied to several different valve types, including butterfly, diaphragm, ball, and plug valves.

Valve body plastics linings are usually limited to those materials having the best all-round chemical resistance properties. The following materials are mainly used:

polypropylene
polyvinylidene fluoride (PVDF)
fluorinated ethylene propylene (FEP)
ethylene tetrafluorethylene (ETFE)
perfluoro alkoxy (PFA)

The fluoropolymers are resistant to a very wide range of chemicals and their upper temperature limit may be as high as 200 °C.

Elastomers are also used to a large extent for valve body linings. These may be moulded in much the same way as plastics but a common method of application, especially for large valves, is to take a sheet of the elastomer and 'tailor' it to the inside of the valve body.

MATERIALS

A wide range of elastomers is used in this way and is most suitable for many chemical services as well as having good abrasion resistance. Among the materials commonly used are:

natural rubbers
styrene rubbers
polyurethane rubbers
ethylene propylene rubbers
butyl rubbers
nitrile rubbers
neoprene rubbers

Plastics and elastomers are used as valve coatings for decorative purposes as well as corrosion resistance. These are applied using spray or dip coating techniques. On external surfaces, epoxy and nylon coatings give good appearance and corrosion resistance. These materials are also applied to internal valve surfaces but for bette corrosion resistance, the fluoropolymer coatings such a PVDF and ethylene chlorotrifluoroethylene (ECTFE are used. Many of the elastomers already described ar applied to valve components in a paint form using suitable solvent.

Although the thickness of these coatings is limited continuous layers are applied and a high degree of corro sion protection is achieved.

It will be seen that there is a large number of plastic and elastomeric materials available for use in valve con struction by the methods described. Although some indi cation of their properties has been given, the suitabilit of a given material for valve service depends upon number of factors. Reference should always be made t manufacturers' literature and recommendations.

TRIM PARTS

The expression 'trim' refers specifically to the disk or disk seats, body seats, and stem, but in some cases it may also include other items such as bushings, bolts and nuts, etc. (Note. The term 'disk' for the closing member of a valve is the standard term used for most types of valves but there are exceptions, e.g., 'plug' in the case of a plug type of valve.)

A brief survey of some of the materials used most frequently for trim parts is given in the following.

Bronze

Used extensively in bronze, iron, and steel valves for general services up to around 280 °C temperature. Appli cations include steam, water, oil, air, and gas lines. It is also used, in the appropriate grade, for disks and body seats (with stainless steel stems) for extremely low tem perature services such as liquid methane, oxygen, and nitrogen.

Zinc-free bronze, usually an aluminium bronze, can also often be provided for specific applications.

Iron

Valves are available with all parts made of iron ('all iron') with the exception of the stem, which is usually of steel. Both disk and body usually have integral seating faces. 'All iron' valves are often an economic choice for handling concentrated sulphuric acid or mixture of the acid with hydrocarbons and are very satisfactory for many other fluids in the chemical and related industries such as brines, ammonia, alcohols, detergents, and chlorinated solvents.

13 Per Cent Chromium Stainless Steel

Used extensively for stems, body seat rings, and disks, this material provides high resistance to wear, galling, corrosion, and erosion with service fluids containing some degree of lubrication. It has high resistance to oxidation and the corrosive action of hot sulphur bearing oils and has served most successfully for many years on oil and vapour lines at temperatures up to 600 °C.

Nickel Alloy

A combination of 'nickel alloy' (in this context, an alloy of nickel, copper, and tin) seat rings and a 13 per cent chrome steel disk is especially suitable for nonlubricating and relatively noncorrosive gases and liquids. Other ser vice applications include superheated or saturated steam, natural gas, fuel oil, gasoline, and low viscosity oils.

Temperature limits are in the order of 450 °C for stean and 260 °C on other services.

A combination of nickel alloy seat rings and disk i also used for steam, water, and other duties.

Austenitic Stainless Steels

Already referred to in connection with valve bod materials, these steels of the basic 18–8 chromium nickel type are also widely used for valve trim part wherever extremes of corrosion and/or temperature are t be met.

Special Stainless Steels

Of these, '20' alloy, 'Incoloy 825', and 'Carpenter 20Cb are the most usually employed for valve trim. They ar used most often in valves made of standard stainles steels but also sometimes in iron or steel valves.

'Monel' Alloy

As a trim material this alloy is mostly used in iron o steel valves for handling sea water, salt solutions, o steam.

'Hastelloy' Alloys 'B' and 'C'

These alloys are not widely used as trim materials being more often used for complete valves. However 'Hastelloy' alloy 'B' is sometimes used as a trim materia in valves handling sulphuric acid or dilute hydrochlori acid, while 'Hastelloy' alloy 'C' is typically used in specia chlorine valves and in valves handling mixed acids.

Hardfacing

Hardfacing involves the deposition of a fused layer o hard material on to disk and body seatings so as to provid high wear resistance and avoid galling, particularly wher elevated temperatures or dry conditions are to be en countered.

Facing materials are usually chosen from cobalt an nickel based alloys and the mating surfaces are generall deposited in similar materials but with a hardness differ ential to reduce the risk of galling in operation. Thi latter condition, however, does not always apply, bein dependent on the properties of the hardfacing alloy used

Cobalt and nickel based hardfacing alloys are availabl from a number of valve manufacturers to very simila specifications and choice will depend on an assessment o many factors. An important consideration is often the relative ease with which a given product can be deposited on a particular component.

Plastics and Elastomers

Plastics and elastomers are used in many different valve types for sealing and seating arrangements in one form or another.

The most widely used plastics material is polytetrafluorethylene (PTFE), which is used for seats in ball and butterfly valves and as a diaphragm facing in diaphragm valves. PTFE is also used for stem packing in many types of valves.

As a seating and sealing material, PTFE is unique; it is resistant to practically all fluids, has very low frictional properties, and can be used at temperatures up to 250 °C. PTFE is often used in its pure form but for many applications, to improve its compression set properties, it is filled with a proportion of relatively inert filler, such as glass.

The chemical resistance of PTFE makes it ideal for use as a diaphragm where its toughness and flex crack resistance are also useful properties. In most valves using a diaphragm, a thin PTFE facing is used in conjunction with an elastomeric backing. The PTFE facing may be separate, or bonded to the elastomer.

A large number of elastomers are used for trim applications in valves in the form of 'O' rings, gaskets, valve seats, and seat inserts, diaphragms, and butterfly valve sleeves.

'O' rings for stem sealing are used in large numbers, a very popular material being nitrile rubber. Fluorocarbon rubbers ('Viton'*), ethylene propylene rubbers,

and silicone rubbers are specified for higher temperature applications.

The spectrum of elastomeric compounds used for valve seats, inserts, diaphragms, and sleeves is extremely wide. The base polymers can be compounded to give a wide range of physical and chemical properties. The most common elastomers for these applications are:

natural rubbers
butyl rubbers
ethylene propylene rubbers
neoprene rubbers
nitrile rubbers
styrene–butadiene rubbers
'Viton'* rubbers

The advantage of using plastics and elastomers for valve trim applications is their resistance to corrosion and erosion and the ability to give a leak-tight seal. Their disadvantage is a limited temperature application, depending upon the material used. Again, the suitability of a given material will depend upon a number of factors. Reference should always be made to manufacturers' literature and recommendations.

Anti-Static and Fire Safe Features

In certain operational circumstances and depending on the valve design there may be a requirement for anti-static devices to be incorporated in valves having non-metallic (soft) seatings so as to ensure electrical continuity between all parts of the valve.

There may also be a requirement for such valves to include 'fire safe' features so that in the event of the soft seats and/or seals being damaged or destroyed by fire the valve will still be operational and any leakage will be within the acceptable limits laid down by some particular standard or specification.

---

* Registered trade mark of E. I. Du Pont de Nemours, Inc.

CORROSION RESISTANCE TABLES

These tables are a summary of the most reliable laboratory data available on the corrosion resistance of the metals commonly used in the construction of valves. It must be noted, however, that the performance of a material in actual plant service can be influenced, sometimes quite significantly, by many factors such as the design and operating characteristics of the valve itself, the type of process involved, fluctuations in concentration and temperature of the medium, presence of other chemicals, and so on. The data shown should, therefore, be used only as a guide to selection of the most likely materials and must not be construed as a guarantee of their suitability. In all cases where there is any doubt further advice should be sought from the valve or material supplier against all available information and corrosion tests should be carried out on site if at all possible.

The following symbols and abbreviations are used in the tables:

SYMBOLS

A   excellent resistance; less than 0·10 mm/year
B   good resistance; 0·10–0·50 mm/year
C   poor resistance; greater than 0·50 mm/year
D   not recommended

ABBREVIATIONS

C20*   'Carpenter' 20Cb-3 stainless alloy
Fr   'Ferralium' alloy
Ic   'Inconel' alloy
Ni   nickel
Ta   tantalum
Ti   titanium
Zr   zirconium
'20' alloy high nickel stainless steel
(20Cr–29Ni–3Mo–3·5Cu–rem. Fe)

'Hastelloy' is a registered trade mark of the Cabot Corporation.

'Monel', 'Inconel', and 'Incoloy' are registered trade marks of Henry Wiggin & Company Limited.
'Ferralium' is a registered trade mark of Langley Alloys Limited.
'Carpenter' 20Cb-3 is a special stainless steel patented and manufactured by Carpenter Technology.

* Other proprietary alloys such as 'Incoloy' alloy '825' may also be considered suitable and enquiries should be made.

Corrosive medium and condition		Temp., °C	Iron and steel	Bronze	Type 304 stainless	Type 316 stainless	'20' alloy	'Monel' alloy	'Hastelloy' alloy 'B'	'Hastelloy' alloy 'C'	Aluminium	Others
acetic acid	5–10%	20	D	D	A	A	A	B	A	A	A	
acetic acid	5–10%	boil	D	D	B	B	A	C	B	A	A	
acetic acid	20%	20	D	D	A	A	A	B	B	A	D	C20: A
acetic acid	50%	20	D	D	A	A	A	B	A	A	D	
acetic acid	80%	20	D	D	A	A	A	B	A	A	A	
acetic acid	80%	boil	D	D	D	B	B	A	B	A	C	
acetic acid	glacial	20	D	D	A	A	A	A	A	A	A	
acetic acid	glacial	boil	D	D	A	B	A	B	A	A	A	Zr: A C20: A
acetic acid vapors	30%	hot	D	D	D	B	B	B	B	A	B	
acetic acid vapors	100%	hot	D	D	D	C	B	B	B	A	B	
acetic anhydride		boil	C	D	B	B	B	B	B	A	B	
acetone		boil	B	A	B	B	A	A	A	A	A	
acetylene		20	A	B	A	A	A	A	A	A	A	
acid mine water		20	D	D	B	B	B	D	C	B	D	
alcohol, ethyl		20	B	B	B	B	A	B	A	A	A	
alcohol, ethyl		boil	B	B	B	B	A	B	A	A	B	
alcohol, methyl		20	B	B	B	B	A	A	A	A	A	
alcohol, methyl		boil	B	B	C	B	A	A	A	A	A	
alum (Al-K-sulphate)	10%	20	D	D	B	B	B	B	B	A	C	
alum	10%	boil	D	D	B	B	A	B	C	B	C	

Corrosive medium and condition		Temp., °C	Iron and steel	Bronze	Type 304 stainless	Type 316 stainless	'20' alloy	'Monel' alloy	'Hastelloy' alloy 'B'	'Hastelloy' alloy 'C'	Aluminium	Others
alum	sat.	boil	D	D	C	B	B	B	C	B	C	
aluminum chloride	25%	20	D	D	D	C	B	B	C	B	D	C20: B
aluminum chloride	25%	boil	D	D	D	D	C	B	C	B	D	Zr: A
aluminum fluoride	5%	20	D	D	D	C	C	A	B	B	D	
aluminum sulphate	all.	20	D	D	B	B	A	B	B	A	B	
aluminum sulphate	all.	boil	D	D	C	B	A	B	C	B	C	
amines		20	A	B	A	A	A	A	A	A	B	
ammonia	all conc.	20	B	D	A	A	A	C	B	A	B	
ammonia gas		hot	C	D	D	D	B	D			D	
ammonium carbonate		20	B	C	B	B	A	B	B	B	A	
ammonium chloride	10%	20	C	D	B	B	A	B	B	B	D	C20: A
ammonium chloride	10%	boil	D	D	C	B	A	B	B	C	D	C20: A
ammonium chloride	25%	boil	D	D	D	C	B	B	B	C	D	
ammonium hydroxide		20	D	B	A	A	A	C	A	A	B	
ammonium hydroxide	conc.	hot	D	C	A	A	A	D	A	A	B	
ammonium nitrate		20	D	A	B	B	A	C	D	A	B	
ammonium nitrate	sat.	boil	D	D	B	B	A	D	B	B	C	
ammonium persulphate	5%	20	D		B	B	A	D	D	A	D	
ammonium monophosphate		20	C	D	A	A	A	B	A	A	D	
ammonium diphosphate		20	D	B	A	A	A	B	A	A	C	
ammonium triphosphate		20	D	B	B	A	A	B	A	A	B	
ammonium sulphate	5%	20	D	A	C	B	B	A	B	B	C	
ammonium sulphate	10%	boil	D	C	D	C	B	B	D	B	D	C20: B
ammonium sulphate	sat.	boil	D	C	D	C	B	B	D	A	D	C20: B
amyl acetate	conc.	20	C	A	B	B	A	A	A	A	B	
amyl alcohol	conc.	20		C			B				D	Cu: B
aniline	3%	20	B	A	A	A	A	B	A	A	B	
aniline	conc.	20	B	A	B	B	A	B	B	B	B	
aniline hydrochloride		20			D	C	C	B	B			
antimony trichloride		20	D	D	D	D		B			D	
aqua regia		20	D	D	D	D	D	D	D	C	D	Ti: B
aqua regia		93	D	D	D	D	D	D	D	D	D	
asphalt chloride	5%	hot		B	B			B			B	
barium chloride	5%	20	C	D	D	B	A	B	B	B	C	
barium chloride	sat.	20	B	D	C	B	A	B	B	B	C	
barium chloride	aq. soln.	hot	D	D	D	C	B	B	B	C	D	
barium sulphate		20	B		B	B	B	B			B	
barium sulphide	sat.	20	D	D	C	B	A	A	A	C	B	
beer		20	B	A	A	A	A	A	A	A	B	
benzene (benzol)		hot	B	A	B	B	A	B	B	B	B	
benzoic acid		20	B	B	B	B	B	B	A	A	B	
bleaching powder		20	D	D	D	B	B	C	D	A	D	
blood (meat juices)		20			B	A	A		A	A	D	
borax	5%	20	B	C	A	A	A	A	A	A	B	
boric acid	5%	hot	D	B	B	B	A	B	A	A	D	
bromine	dry	20	C	D	D	D	D	B	B	B	D	Zr: A
bromine water		20	D	D	D	D	D	D	D	B	D	
buttermilk		20	C	A	A	D	A	A	A	A	D	
butane		20		B		B	A	B			B	
butyl acetate		20	A	A	B	B	B	A	A	A	A	
butyric acid	5%	66	B	A	B	B	B	B	B	A	C	
butyric acid	aq. soln.	boil	D	B	B	B	B	B	D	A	D	
calcium bisulphite		20	D	C	C	B	B	B	C	B	D	
calcium carbonate		20	B	C	B	B	B	B	B	B	C	
calcium chloride	dil.	20	D	C	C	B	A	A	B	A	C	C20: A
calcium chloride	conc.	20	D	C	D	C	B	A	B	A	C	
calcium chloride	conc.	boil	D	D	D	D	B	A	A	A	D	
calcium hydroxide	5%	20	C	D	B	B	A	A	A	A	D	
calcium hydroxide	10%	boil	D	D	B	B	B	A	A	A	D	
calcium hydroxide	20%	boil	D	D	D	B	B	A	A	A	D	
calcium hydroxide	50%	boil	D	D	D	B		A	A	A	D	C20: B
calcium hypochlorite	2%	20	D	D	D	C	B	C	C	B	D	
calcium sulphate	sat.	20	B	B	C	B	B	B	B	B	B	
carbolic acid	c.p.	boil	C	C	B	B	B	B	B	B	C	
carbolic acid		boil	C	D	B	B	B	B	B	B	C	

Corrosive medium and condition		Temp., °C	Iron and steel	Bronze	Type 304 stainless	Type 316 stainless	'20' alloy	'Monel' alloy	'Hastelloy' alloy 'B'	'Hastelloy' alloy 'C'	Aluminium	Others
carbon disulphide		20	B	D	B	B	B	C	B	B	B	
carbon monoxide		200	A	B	A	A	A	A	A	A	B	
carbon monoxide		820	D	D	B	A	A	C	A	A	D	
carbon tetrachloride	c.p.	20	B	B	B	B	A	A	B	A	B	
carbon tetrachloride	c.p.	boil	C	C	C	B	A	A	B	A	D	
carbon tetrachloride		boil	D	D	C	C	A	B	B	A	D	
carbonic acid	sat.	20	D	C	B	B	A	B	A	A	B	
chloric acid		20	D	D	D	D	B	D	D	A	D	
chlorinated water	sat.	20	D	D	D	C	C	D	D	B	D	
chlorine gas	dry	20	B	C	D	C	A	B	B	A	B	
chlorine gas	wet	100	D	D	D	D	D	D	D	D	D	Ta: B
chloroacetic acid		20	D	D	D	D	D	D	B	B	D	
chlorobenzene	conc.	20	A	A	A	A	A	A	A	A		
chloroform		20	A	A	A	A	A	A	B	B	A	
chlorosulphonic acid	10%	20	D	D	C	C	C	A	A	B		
chlorosulphonic acid	conc.	20	D	D	B	B	B	D	A	A	B	Zr: A
chromic acid	5%	20	B	C	B	B	B	B	D	B	B	
chromic acid	10% c.p.	boil	D	D	C	C	C	D	D	B	B	
chromic acid	50% com.	boil	D	D	D	D	D	D	D	B	D	
citric acid	5%	20	D	C	A	A	A	A	A	A	B	
citric acid	5%	66	D	C	B	B	A	B	A	A	B	
citric acid	15%	boil	D	D	B	B	A	B	A	A	C	Ni: A
citric acid	conc.	boil	D	D	D	B	B	B	A	A	B	
copper acetate	sat.	20	D	D	B	B	B	B	A	A	D	
copper carbonate	sat.	20			A	A	B	A	A	A		
copper chloride	1%	20	D	D	C	C	B	D	C	B	D	C20: A
copper chloride	5%	20	D	D	D	C	B	D	C	B	D	C20: B
copper chloride	5%	boil	D	D	D	D	C	D	D	C	D	Ti: B
copper cyanide	sat.	boil			B	B	B	C	B	B		
copper nitrate	5%	20	D		A	A	A	A	D	B		
copper nitrate	50%	hot	D		B	B	B	B	D	B		
copper sulphate	5%	20	D	D	B	B	A	D	D	B	D	C20: A
copper sulphate	sat.	boil	D	D	B	B	A	D	D	B	D	C20: A
creosote (coal tar)		hot	B	C	B	A	B	A	A	A	B	
cresylic acid		20	B	B	A	A	A	A	A	A	A	
developing solutions		20	D		B	B	A	B	A	A		
dichloreothane		boil	D		B	B	B	B	B	B		
'Dowtherm A'		boil	A	D	A	A			A	A	C	
ether		20	A	A	A	A	A	A	B	A	A	
ethyl acetate	conc.	20	B	A	A	A	A	B	B	B	A	
ethyl chloride	dry	20	A	A	A	A	A	B	B	B	A	
ethylene glycol		20	B	A	A	A	A	A	A	A	A	
fatty acids		boil	C	B	B	B	A	A	B	A	C	Ic: A
ferric chloride	1%	20	D	D	D	C	C	D	D	B	D	C20: A
ferric chloride	1%	boil	D	D	D	D	D	D	D	C	D	Ti, Zr: B
ferric chloride	5%	20	D	D	D	D	D	D	D	B	D	
ferric hydroxide		20			A	A	A	A	A	A		
ferric nitrate	5%	20			B	B	A	B	C	B	D	
ferric sulphate	5%	20	D	D	B	A	A	C	D	B	D	
ferric sulphate	5%	boil	D	D	B	B	A	D	D	B	D	
ferrous sulphate	10%	20	D	B	B	B	B	B	B	B	C	
ferrous sulphate	sat.	20	D	D	B	B	B	B	B	B	D	
fluorine	dry	20	C	B	B	B	A	B	B	B	D	
fluosilicic acid		20	D	D	D	D	B	B	B	B	D	C20: B
formaldehyde		20	B	B	B	B	A	B	B	B	B	
formic acid	5%	20	C	C	B	B	A	B	C	A	D	
formic acid	5%	66	D	C	B	B	B	C	C	A	D	
formic acid	10–50%	20	C	C	B	B	A	B	B	B	D	
formic acid	10–50%	boil	D	D	D	D	B	C	C	B	D	C20: B
formic acid	100%	20	D	C	C	C	A	B	B	A	D	C20: A
formic acid	100%	boil	D	C	D	D	B	C	C	B	D	C20: B
freon	dry		A	A	A	A	A	A	A	A	A	
freon	wet		B	B	C	C	C	B	B	B	C	
fruit juices		20	C	A	A	A	A	B	A	A	D	Ic: A
fuel oil		hot	B	A	A	A	A	A	A	A	A	

Corrosive medium and condition		Temp., °C	Iron and steel	Bronze	Type 304 stainless	Type 316 stainless	'20' alloy	'Monel' alloy	'Hastelloy' alloy 'B'	'Hastelloy' alloy 'C'	Aluminium	Others
furfural		20	B	B	B	B	A	B	B	B	B	
gallic acid	5%	20	B	C	B	B	A	B	B	B	D	
gallic acid	5%	66	C	D	B	B	B	B	B	B	D	
gasoline	refined	20	A	A	A	A	A	A	A	A	A	
gasoline	sour	20	D	A	A	A	A	D	A	A	C	
gelatine			D	A	A	A	A	A	A	A	D	
glucose				B	B	B	A	B			B	
glycerine		20	B	A	A	A	A	A	A	A	A	
hydrobromic acid		20	D	D	D	D	D	C	B	C	D	
hydrocarbons (aliphatic)		20	A	A	A	A	A	A	A	A	A	
hydrocarbons (aromatic)		20	A	A	A	A	A	A	A	A	A	
hydrochloric acid	1%	20	D	D	D	C	C	C	A	A	D	C20: A
hydrochloric acid	1%	boil	D	D	D	D	D	D	B	D	D	Zr: A
hydrochloric acid	5%	20	D	D	D	D	D	D	A	B	D	C20: B
hydrochloric acid	5%	boil	D	D	D	D	D	D	B	D	D	Zr: A
hydrochloric acid	10%	20	D	D	D	D	D	D	A	C	D	Zr: A
hydrochloric acid	10%	boil	D	D	D	D	D	D	B	D	D	
hydrochloric acid	25%	20	D	D	D	D	D	D	A	B	D	Zr: A
hydrochloric acid	25%	boil	D	D	D	D	D	D	B	D	D	
hydrochloric acid	conc.	20	D	D	D	D	D	D	A	B	D	
hydrochloric acid	conc.	boil	D	D	D	D	D	D	B	D	D	Zr: B
hydrocyanic acid		20	D	A	B	B	B	B	B	B	C	
hydrofluoric acid	conc.	20	D	D	D	D	C	A	B	B	D	C20: B
hydrofluoric acid	conc.	80	D	D	D	D	D	B	B	B	D	C20: B
hydrofluorosilicic acid		20	D	D	D	D	B	B	B	B	D	
hydrogen chloride	dry	20	C	D	C	B	A	B	B	B	C	
hydrogen chloride	wet	20	D	D	D	C	C	C	B	A	D	
hydrogen fluoride	dry	20	C	D	C	C	A	B	B	B	D	
hydrogen fluoride	wet	20	D	D	D	D	C	B	B	B	A	
hydrogen peroxide		20	D	C	A	A	A	B	B	A	A	C20: B
hydrogen peroxide		boil	D	D	B	B	B	C	C	B	B	
hydrogen sulphide	dry	20	A	B	A	A	B	B	B	A	A	Ic: A
hydrogen sulphide	wet	20	C	C	C	B	C	C	C	B	B	
hydriodic acid	dil.	20	D	D	D	D		C	C	C	D	
iodine	dry	20	D	D	D	C	B	C	C	B	D	
iodine	wet	20	D	D	D	D	D	D	D	B	D	
idoform		20	D		A	A	A	A	A	A		
kerosene		20	A	A	A	A	A	A	A	A	A	
ketchup		20	D	C	B	A	A	B	A	A	A	
lacquers		hot	D	B	B	B	A	B	B	B	B	
lactic acid	1%	boil	D	D	B	B	B	C	B	C	D	
lactic acid	5%	20	D	D	B	A	A	B	B	B	A	
lactic acid	5%	66	D	D	B	B	A	C	B	B	D	
lactic acid	5%	boil	D	D	D	B	B	D	B	C	D	
lactic acid	10%	20	D	D	B	A	A	B	B	B	A	
lactic acid	10%	66	D	D	C	B	A	C	B	B	D	C20: B
lactic acid	10%	boil	D	D	D	B	B	D	B	C	D	
lactic acid	conc.	20	D	D	B	A	B	B	B	B	D	C20: B
lactic acid	conc.	boil	D	D	D	D	B	D	B	C	D	
lead acetate		20	D		B	B	B	B	B	B	D	
lithium		150	B	D	A	A		C	C	C	D	Ta: A
lithium		540	C	D	C	C		C	C	C	D	Ta: A
lithium		820	C	D	C	C		D	D	D	D	Ta: B
lye (caustic)		20	D	D	B	B	B	A	B	B	D	
lye (caustic)		boil	D	D	B	B	B	A	B	B	D	
'Lysol'		20	D	D	C	C		D	C	B	D	
magnesium carbonate		20			B	B	B	B	B	B		
magnesium chloride	5%	20	D	C	B	B	A	A	A	A	C	
magnesium chloride	5%	hot	D	D	D	D	A	A	A	B	D	
magnesium chloride	10–30%	20	D	D	C	B	A	A	A	A	C	
magnesium chloride	sat.	20	D	D	C	B	B	A	A	A	C	
magnesium hydroxide		20	B	B	B	B	B	A	A	B	D	
magnesium sulphate		20	B	B	B	B	A	A	A	A	B	
magnesium sulphate		hot	C	B	B	B	A	A	A	B	B	
malic acid	conc.	20	D		B	B	B	B	B	B	B	

Corrosive medium and condition		Temp., °C	Iron and steel	Bronze	Type 304 stainless	Type 316 stainless	'20' alloy	'Monel' alloy	'Hastelloy' alloy 'B'	'Hastelloy' alloy 'C'	Aluminium	Others
malic acid		hot	D		B	B	B	B	B	B	C	
mercuric chloride	2%	20	D	D	D	D		D	D	B	D	C20: A
mercuric chloride	2%	hot	D	D	D	D		D	D	B	D	C20: B
mercuric cyanide		20	D		B	B	B		B	B		
mercurous nitrate		20	D		B	B	B		D	B		
mercury		150	B	C	C	C		C			D	
mercury		540	D	D	D	D		D			D	
mercury		820	D	D	D	D		D			D	
methyl chloride gas		20	D	B	B	B	A	B	B	B	D	
milk		20	D	C	A	A	A	B	A	A	A	
mixed acids 1% sulphic, 99% nit.		20	D	D	B	B	B	D	D	B	B	
mixed acids 1% sulphic, 99% nit.		110	D	D	C	C	C	D	D	C	D	
mixed acids 10% sulphic, 90% nit.		20	D	D	B	B	B	D	D	B	B	Ta: B
mixed acids 10% sulphic, 90% nit.		110	D	D	C	C	C	D	D	C	D	
mixed acids 15% sulphic, 5% nit.		20	D	D	B	B	A	D	D	B	B	Ta: B
mixed acids 15% sulphic, 5% nit.		110	D	D	C	C	B	D	D	C	D	Ta: B
mixed acids 30% sulphic, 5% nit.		20	D	D	B	B	A	D	D	B	C	
mixed acids 30% sulphic, 5% nit.		110	D	D	C	C	B	D	D	C	D	Ta: B
mixed acids, 53% sulphic, 45% nit.		20	B	D	B	B	B	D	D	B	D	
mixed acids, 53% sulphic, 45% nit.		110	D	D	C	C	B	D	D	C	D	Ta: B
mixed acids, 58% sulphic, 40% nit.		20	B	D	B	B	B	D	D	B	D	
mixed acids, 58% sulphic, 40% nit.		110	D	D	D	D	C	D	D	C	D	Ta: B
mixed acids, 70% sulphic, 10% nit.		20	D	D	C	C	B	D	D	B	D	
mixed acids, 70% sulphic, 10% nit.		110	D	D	D	D	C	D	D	D	D	Ta: B
muriatic acid		20	D	D	D	C	B	C	B	A	D	
naphtha		20	B	B	B	B	B	B	B	B	B	
nickel chloride		20	D	D	C	B	B	B	A	A	D	
nickel sulphate		hot	D	D	C	B	B	B	D	B	D	
nitric acid	1%	20	D	D	A	A	A	D	D	A	B	
nitric acid	1%	hot	D	D	B	B	A	D	D	C	D	
nitric acid	5%	20	D	D	A	A	A	D	D	A	B	
nitric acid	10%	20	D	D	A	A	A	D	D	A	D	
nitric acid	10%	boil	D	D	B	C	B	D	D	C	D	
nitric acid	20%	20	D	D	A	A	A	D	D	B	D	
nitric acid	50%	20	D	D	A	A	A	D	D	B	C	
nitric acid	50%	boil	D	D	B	C	B	D	D	D	D	Zr: A, Ti: B
nitric acid	65%	boil	D	D	B	B	B	D	D	D	C	
nitric acid	85%	20	B	D	B	A	B	D	D	D	B	
nitric acid	85%	hot	D	D	B	C	B	D	D	D	D	Zr: A
nitric acid	conc.	20	B	D	B	C	B	D	D	B	A	
nitric acid	conc.	hot	D	D	C	C	B	D	D	D	D	Zr: A
nitrous acid	5%	20	D	D	B	B	B	D	D	B	C	
oils (crude)		hot	D	C	B	B	A	B	B	B	B	
oils (veg. & mineral)		hot	B	C	B	B	A	B	B	B	B	
oleic acid		20	C	C	B	B	B	A	B	B	B	
oleic acid	raw	205	D	D	C	B	B	B	B	B	D	
oleum		20	B	D	C	B	A	D	B	A	C	
oleum		hot	D	D	D	C	B	D	B	A	D	
oxalic acid	5%	hot	D	D	C	C	B	D	C	B	D	
oxalic acid	10%	20	C	C	B	B	B	B	B	B	D	C20: A
oxalic acid	10%	boil	D	D	D	D	B	C	B	B	D	C20: B
oxalic acid	50%	boil	D	D	D	D	B	C	B	B	D	
oxalic acid	sat.	20	C	D	B	B	B	B	B	B	D	
oxalic acid	sat.	boil	D	D	D	D	B	C	B		D	
oxygen		cold	B		A	A	A	A			A	
oxygen		260			B	B	A	B		A	B	
oxygen		260–540			B	B	B	B		C	C	
oxygen		540	D		D	C	B	D		C	C	347: B
palmitic acid		20	C	B	B	B	B	B	B	B	D	
paraffin		hot	A	A	A	A	A	A	A	A	A	
phenol	c.p.	boil	C	C	B	B	B	B	B	B	B	
phosphoric acid, c.p.	1%	20	D	B	B	B	A	B	A	A	B	
phosphoric acid	5%	20	D	C	B	B	A	B	A	A	B	
phosphoric acid	10%	20	D	C	B	B	A	B	A	A	C	
phosphoric acid	10%	boil	D	D	D	C	B	D	A	A	D	C20: A

Corrosive medium and condition		Temp., °C	Iron and steel	Bronze	Type 304 stainless	Type 316 stainless	'20' alloy	'Monel' alloy	'Hastelloy' alloy 'B'	'Hastelloy' alloy 'C'	Aluminium	Others
phosphoric acid	25%	boil	D	D	D	C	B	D	A	B	D	C20: A Fr: A
phosphoric acid	45%	70	D	D	D	B	A	B	A	A	D	Fr: A
phosphoric acid	45%	boil	D	D	D	C	B	D	A	B	D	C20: A Fr: A
phosphoric acid	85%	20	D	D	D	B	A	D	A	A	D	Fr: A
phosphoric acid	85%	boil	D	D	D	D	C	C	A	C	D	C20: B
picric acid, aqueous soln		20	C	D	B	B	B	D	B	B	D	
potassium bromide		20	D	B	B	C	B	B	B	B	C	
potassium carbonate	1%	20	B	B	B	B	A	B	B	B	B	
potassium chlorate		20	B	B	B	B	A	B	C	B	B	
potassium chloride	1–5%	20	D	C	C	B	A	A	B	A	C	
potassium chloride	1–5%	boil	D	D	D	D	A	B	B	B	D	C20: A
potassium cyanide		20	B	D	B	B	A	B	B	B	B	
potassium dichromate		20	C	C	B	B	A	C	C	B	B	
potassium ferricyanide	5%	20	C	D	B	B	B	B	B	B	B	
potassium ferrocyanide	5%	20	C	D	B	B	B	B	B	B	B	
potassium hydroxide	5%	20	B	D	B	B	B	A	B	B	D	C20: A
potassium hydroxide	25%	boil	D	D	B	B	B	A	B	B	D	C20: A
potassium hydroxide	50%	boil	D	D	B	B	B	A	B	B	D	
potassium hypochlorite		20	D	D	D	C	B	D	D	B	D	
potassium nitrate	1–5%	20	B	B	B	B	B	B	C	B	B	
potassium nitrate	1–5%	hot	B	C	B	B	B	B	C	B	B	
potassium permanganate	5%	20	A	A	A	A	A	A	A	A	A	
potassium sulphate	1–5%	20	B	B	B	B	A	B	B	B	B	
potassium sulphate	1–5%	hot	D	B	B	B	B	B	B	B	B	
potassium sulphide	sat.	20	C	D	B	B	B	B	B	B	D	
propane			B		B	B	A	B			B	
propionic acid		20	D	D	B	B	A	A	A	A	B	
pyrogallic acid		20	B	B	B	B	B	B	B	B	B	
quinine bisulphate	dry	20	D	B	B	B	B	B	B	B	D	
quinine sulphate	dry	20	D	B	B	B	B	B	B	B	D	
rosin	molten		D	C	A	A	A	A	A	A	A	
sal ammoniac		20	C	D	B	B	B	B	B	B	D	
salicylic acid		20	D	B	B	B	B	B	B	B	B	
sea water		20	D	C	B	B	A	A	A	A	C	
sewage		20	C	C	B	B	A	C	B	B	C	
silver bromide		20	D		C	B	A		B	B		
silver chloride		20	D		D	D	C		D	B		
silver nitrate		20			B	B	A	D	B	B		
soaps		20	B	B	B	B	A	B	B	B	B	
sodium and/or sodium–potassium NaK		150	B	C	A	A	A	A	A	A	C	Zr: A
sodium and/or sodium–potassium NaK		540	D	D	A	A	A	B	A	A	D	Ic, Ni, Ti, Ta-A
sodium and/or sodium–potassium NaK		820	D	D	A	A	A	D	A	A	D	
sodium acetate	moist	20	C	B	B	B	A	B	B	B		
sodium bicarbonate		20	C	B	B	B	A	B	B	B	C	
sodium bisulphate		20	D	C	B	B	B	B	B	B	D	
sodium carbonate	5%	20	B	B	B	B	A	B	B	B	D	
sodium carbonate	5%	66	B	D	B	B	A	B	B	B	D	
sodium chlorate	10%	20			B	B	B		D	B	B	
sodium chlorate	25%	20			B	B	B	A	D	B	B	
sodium chloride	5%	20	C	B	B	B	A	A	B	B	C	
sodium chloride	20%	20	C	B	B	B	A	A	B	B	C	
sodium chloride	sat.	20	C	B	B	B	A	A	B	B	C	
sodium chloride	sat.	boil	D	D	C	B	B	A	B	B	D	
sodium cyanide		20	B	D	B	B	A	A	B	B		
sodium fluoride	5%	20			B	B	B	B	B	B		
sodium hydroxide	5%	20	B	C	B	B	A	A	A	B	D	C20: A
sodium hydroxide	20%	boil	B	D	B	B	B	A	A	B	D	C20: A
sodium hydroxide	50%	boil	D	D	B	B	A	A	A	B	D	C20: A
sodium hydroxide	75%	boil	D	D	D	C	B	A	B	B	D	Ti: A
sodium hypochlorite	5%	20	D	D	D	C	C	C	C	B	D	
sodium hyposulphite		20	D	D	B	B	B	B	B	B	D	
sodium nitrate		20	B	B	A	A	A	B	D	B	B	
sodium silicate			B	B	B	B	A	B	B	B	C	
sodium sulphate	sat.	20	B	B	C	C	A	B	B	B	C	
sodium sulphide	sat.	20	B	D	C	B	A	B	B	B	D	

Corrosive medium and condition		Temp., °C	Iron and steel	Bronze	Type 304 stainless	Type 316 stainless	'20' alloy	'Monel' alloy	'Hastelloy' alloy 'B'	'Hastelloy' alloy 'C'	Aluminium	Others
sodium sulphite	5%	20	B	D	B	B	A	B	D	B	B	
sodium sulphite	10%	66	B	D	C	B	A	B	D	B	B	
sodium sulphite	10%	boil	D	D	C	B	B	B	D	B	C	
sodium thiosulphate	20%	20	D	D	B	B	B	B	B	B	B	
stannic chloride	5%	20	D	D	D	C	B	D	B	B	D	
stannic chloride	5%	boil	D	D	D	D	D	D	C	B	D	
stannous chloride	sat.	20	D	D	D	C	B	C	B	B	D	Zr: A
steam		100	A	B	A	A	A	A	A	A	B	
steam		205	A	B	A	A	A	A	A	A	B	
steam		320	C	D	A	A	A	A	A	A	D	
stearic acid		20	C	C	B	A	A	A	A	A	A	
strontium nitrate		20	D	C	A	A	A	A	A	A	A	
sugar juice		66	D	B	B	B	A	B	B	B	B	
sulphate liquors		20	C	D	B	B	B	B	B	B	D	C20: A
sulphite liquors		20	D	D	C	B	B	D	C	B	C	C20: A
sulphur—dry, molten			B	C	B	B	B	B	B	B	B	
sulphur—wet, molten			D	D	C	B	B	C		B	C	
sulphur chloride			D	D	C	B	B	B	D	B	B	
sulphur dioxide	dry	260	B	D	B	B	B	B	C	B	B	
sulphur dioxide	moist	20	D	D	C	B	B	C	C	B	B	
sulphuric acid	1%	20	D	B	B	B	A	C	A	A	B	
sulphuric acid	1%	boil	D	D	C	C	B	B	B	C	C	C20: B; Zr: A
sulphuric acid	5%	20	D	C	C	B	A	C	A	A	C	
sulphuric acid	5%	boil	D	D	D	C	B	B	B	C	D	C20: B; Zr: A
sulphuric acid	10%	20	D	C	D	C	B	C	A	A	C	
sulphuric acid	10%	boil	D	D	D	D	B	B	B	C	D	Zr: A
sulphuric acid	50%	20	D	D	D	D	A	B	A	A	D	C20: A
sulphuric acid	50%	boil	D	D	D	D	C	D	A	D	D	
sulphuric acid	60%	20	D	D	D	D	A	B	A	A	D	C20: A
sulphuric acid	60%	boil	D	D	D	D	D	D	C	D	D	Zr, Ta: B
sulphuric acid	80%	20	D	D	C	C	B	C	A	A	D	
sulphuric acid	80%	boil	D	D	D	D	D	D	D	D	D	Ta: B
sulphuric acid	conc.	20	B	D	B	B	A	D	A	A	B	
sulphuric acid	conc.	boil	D	D	D	D	D	D	D	D	D	Ta: 3
sulphuric acid	conc.	20	D	D	D	D	D	D	D	D	D	Ta: B
sulphuric acid	fuming	20	C	D	C	B	A	D	A	A	A	
sulphurous acid	sat.	20	D	D	D	B	B	D	D	B	C	C20: B
sulphurous acid	sat.	120	D	D	D	B	B	D	D	B	D	C20: B
sulphurous spray		20	D	D	D	D		D	D	B	D	
tannic acid		20	D	B	B	B	A	B	B	B	C	C20: A
tannic acid			D	C	C	B	A	B	B	B	D	C20: A
tartaric acid	10%	20	D	C	A	A	A	A	B	B	A	
tartaric acid	10%	hot	D	D	C	B	A	B	B	B	D	C20: A
titanium tetrachloride	wet	20	D	D	D	D	D	B	B	B	D	
trichloroethylene	dry	20	B	B	B	B	B	B	B	A	B	
trichloroacetic acid		20	D	D	D	D	D	D	B	B	D	Zr: A
tri-sodium phosphate			B	D	B	B	B	B	B		D	
turpentine		20	A	B	A	A	A	A	A	A	D	
uric acid	conc.	20	D	D	B	B	B		B	B	A	
varnish		20	C	B	A	A	A	A	A	A	A	
varnish		hot	D	D	B	B	A	A	A	A	B	
vegetable juices		boil	D	D	B	A	A	B	B	A	B	
vinegar		20	D	C	A	A	A	A	A	A	B	
vinegar		hot	D	D	B	B	B	B	B	A	C	
vinegar fumes			D	D	B	B	B	C	B	A	D	
water, distilled			D	D	A	A	A	B	A	A	A	
whiskey and wine		70	D	B	A	A	A	B	A	A	C	
zinc chloride	5%	70	D	D	C	B	A	B	B	B	B	
zinc chloride	5%	boil	D	D	D	C	B	B	B	C	D	
zinc chloride	20%	20	D	D	D	B	B	B	B	B	C	Ti, Zr: A
zinc chloride	20%	boil	D	D	D	C	B	C	B	C	D	C20: B; Zr: A
zinc sulphate	5%	20	B	B	B	A	A	A	B	B	B	
zinc sulphate	25%	boil	D	D	C	B	B	A	B	B	C	
zinc sulphate	sat.	20	B	C	A	A	A	A	B	B	B	

STANDARDS

An important aspect of valves is the wide spread of Standards applying to their construction and performance. At least, this is helpful to the user in that in many cases it results in valves of similar basic design and comparable minimum performance being available from different manufacturers to the same dimensions so that they are interchangeable without difficulty. At most, in some countries or in certain circumstances, the Standards are legally binding.

It is beyond the scope of this book to detail all Standards or to indicate their status in respect of legislation in all countries, not least because such aspects are subject to continual change. The objective here is to indicate the background to the subject of standardisation so that the reader can more readily establish for himself the circumstances in which he is operating.

Most industrialised countries have centralised, national or nationally recognised authorities for the preparation of Standards. These authorities usually bring together users, manufacturers and other interests for the purpose of preparing national Standards for materials, components and finished products made by industry and used either by industry or consumers. The national authority in the United Kingdom is the British Standards Institution.

In turn these national bodies combine to produce common Standards for international acceptance. The Western European body is the European Committee for Standardisation (CEN) and the world body is the International Standards Organisation (ISO).

These international bodies have done much excellent work and continue to do so but at the time of going to press, there is still a very great deal to be done to complete the range of Standards available and then to gain acceptance of them by users and manufacturers in the different countries. The task which has been embarked upon by ISO is a daunting one and will take a very long time to complete. The ultimate advantage to the users and to international trade is, however, too great for the programme not to be expected to come to realisation in the fullness of time.

Whilst the national and international authorities recognise the need for particular Standards and for their regular review and revision, the need for some Standards or changes often becomes apparent at a rather lower level such as in technical trade bodies and by manufacturers associations. These bodies may then draft particular Standards or aspects of them and submit them to their national authority. In other cases such 'trade' Standards may first be adopted within the narrower limits of a particular industry as a recommendation or Code of Practice and then be refined and modified in the light of experience before being submitted for adoption as a true Standard.

Such documents may become quite widespread in use before becoming true Standards. An example is the numerous such items prepared by the American Petroleum industry.

The principal Standards for valves are concerned with basic design details, materials of construction, some dimensions—specific for assembly into pipework and minima for some components—marking, testing and quality control.

There are also Standards for flanges, threaded connections and for pipes and vessels which will, or may have, a bearing on the actual specification of the valve to be employed.

For details of specific Standards applying to valves of a particular type for a particular industry or purpose at any time you should consult your national Standards authority and the technical section of your industry or trade association.

ACCESSORIES

Draining or Purging Point

On valves of most types, especially on the larger sizes, provision is usually made for draining or purging. Commonly this is in the form of a boss or bosses located at strategic points on the valves.

The bosses may be left blind for finishing on site to suit the particular operating conditions or may be tapped and fitted with headed plugs before despatch, depending on customers' requirements.

Position Indicator

On valves where the stem rises, e.g. certain kinds of gate, globe, and diaphragm valves, the position of the stem may often serve to indicate the degree of opening of the valve. Alternatively, and for other valves of these types with nonrising stems, an open–shut indicator can usually be provided, commonly in the form of a pointer travelling up and down a fixed scale plate.

For quarter turn valves, such as ball, butterfly, and plug valves, the position of the operating lever or tommy bar usually serves to indicate the state of the valve opening. Many such valves are also marked on the head to indicate the position of the disk or plug.

Locking Device

A means of securing the valve handwheel or lever to some fixed part of the valve and thus prevent it from being operated. The simplest and most common method is to use a padlock or padlock and chain.

Bypasses

These may be required for such duties as warming up on steam line services and for equalizing the pressure around the disk of the main valve before operating. The bypass valve may be attached to the main valve by piping, or may form an integral part of the main valve if the type and size of the main valve makes this possible.

Chainwheel Operation

Convenient for operating valves located above normal reach from floor levels. The chainwheel may be fitted in place of the usual handwheel, as depicted in Fig. 1, or, in another arrangement, it can be attached to the valve handwheel. Guides hold the chain in close contact with a large portion of the circumference of the chainwheel to prevent slipping or jumping of the chain.

Extension Controls

Extension stems are frequently provided for valves that are to be operated from a distance. They may also sometimes be used to allow room for the fitting of insulation material, or lagging, around the valve without fouling the operating mechanism.

Long extension stems must be properly supported and floorstands are often used for this purpose, as illustrated in Fig. 2. Gearing can also be incorporated for cases where the extended control is not in line with the valve stem.

Reduction Gearing and Gear Operators

Larger sizes of valves, or valves used for unusually high pressures, frequently use reduction gearing to facilitate operation.

The gearing, with its mountings etc., may form an integral part of the valve construction. More usually it is provided in the form of a separate purpose designed unit, suitable for bolting to a prepared mounting on the valve. Such units are available in a variety of designs and size ranges to meet the requirements of practically every type of valve application. An example is shown in Fig. 3.

Fig. 1. Chainwheel in place of handwheel

Fig. 2. Gate valve with extension stem and floorstand

Fig. 3. Bevel gear manual operator

POWER OPERATION

The operation of a valve by means of some form of power device may be required for several reasons, including the following:

(1) The force required to operate the valve is too great for convenient manual operation.

(2) A faster operation is desired than can be achieved by manual means.

(3) It is required to operate the valve remotely.

(4) Operation of the valve is to be in a controlled sequence related to other equipment functions.

(Note. Automatic process control valves form a special category of power controlled valves and are described in a separate section.)

To meet these various operational requirements and to cope with the great variety of valve types and wide range of sizes of valves used in modern industry, many different types of valve actuators are produced, offering the user a wide choice of design features and performance characteristics.

Earlier valve actuators were usually built-in to form an integral unit with the valve, but present day actuators are developed and produced as separate units by specialists and purchased and fitted by the valve user or valve manufacturer. The latter arrangement is the more satisfactory and most valve manufacturers are pleased to provide this service.

Basically, valve actuators may be classified according to the form of power used, that is, pneumatic, hydraulic, electric, or a combination of any two of these, e.g., electro-hydraulic. Selection for a particular application will depend on the power source available, compatibility with the type of valve to be operated, the function or functions required of the actuator, and, maybe, the space available for the actuator.

The accompanying illustrations (Figs 1 to 5) show some of the various types of valve actuators in regular use today. It should however be borne in mind that, because of the large number of producers of actuators and the many different designs available, these illustrations must be purely representative.

Brief references to the basic types of actuators and some of their features appear in the section on automatic process control valves. For complete information, and when considering the selection of an actuator for a specific application, reference should be made to manufacturers, whose specialized assistance is readily available.

Fig. 1. Pneumatic, semi-rotary, vane actuator mounted on a plug valve

Fig. 2. Hydraulic, double acting, semi-rotary actuator applied to a butterfly valve

Fig. 3. Hydraulic, linear type, actuator applied to a gate valve

Fig. 4. Electric actuator applied to a butterfly valve

Fig. 5. Electric actuator mounted on a gate valve

VALVE SELECTION

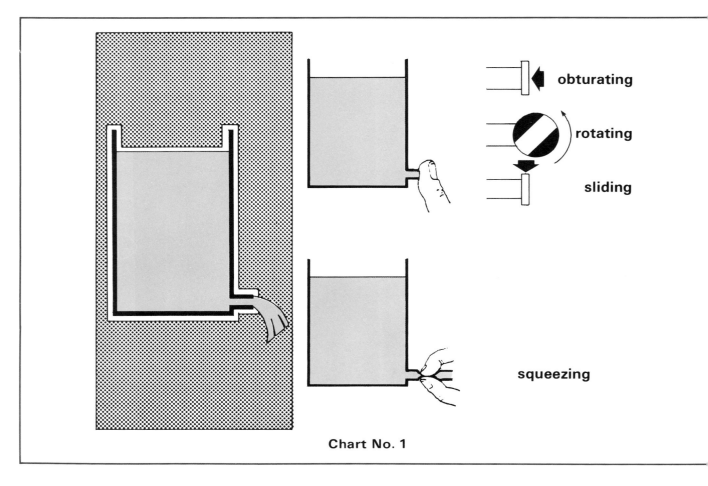

Chart No. 1

CHART 1

Fundamentally there are only two known ways of controlling the flow of liquids and gases, and all valves are in essence based on one or the other of the two principles illustrated in Chart 1. This illustrates a simple tank with liquid flowing from an outlet near the bottom. To stop or control the flow one can either place a finger against the pipe end (Dutch boy) or alternatively, if the pipe is flexible, it can be squeezed (tourniquet). These are the two basic principles on which valves are constructed.

The first principle is developed in three ways:
(1) Moving the stopper by direct thrust on to the orifice seating. This obturating movement is the basis of globe type valves.
(2) Rotating the stopper. This is the basis for plug type valves.
(3) Sliding the stopper across the face of the orifice seating. This is the basis of gate type valves.

The second principle, squeezing action, is the basis of all diaphragm type valves.

CHART 2

Each of the four valve motivations – obturating, rotating, sliding and squeezing – has its own advantages and disadvantages, as listed in Chart 2.

To summarize, there are four basic valve types, each type having its own individual characteristics and offering its own particular advantages or disadvantages. For this reason, many modifications and variations of the basic types are made by valve manufacturers. An appraisal of the chief characteristics of the four basic valve types is given in the following.

		ADVANTAGES	DISADVANTAGES
GLOBE 1		Best shut-off and regulating characteristics	High head loss
PLUG CONICAL		Quick acting. Straight through flow	Temperature limitations on PTFE sleeved valves and need for attention to 'lubricant' in lubricated valves
2 BALL		Quick acting. Straight through flow. Easy operation	Temperature limited by seating material
BUTTERFLY		Quick acting. Good regulating characteristics Compact	Metal to metal seated type does not give tight shut-off. Temperature limited by seating material on resilient seated type
GATE 3		Straight through flow	Slow acting. Bulky
DIAPHRAGM 4		Glandless. Positive shut-off on dirty fluids	Pressure and temperature limited by diaphragm material

Chart No. 2

GLOBE VALVE

Here the direct thrust of the disk on to the seating provides the best form of shut-off and the best regulating characteristics. Closure is positive and it is possible to 'feel' when the valve is shut. The regulating characteristic of a valve is the relationship between movement of the valve handwheel and the effect that this movement has on the amount of medium flowing through the valve.

The globe valve is the most suitable type for throttling, that is, fine regulation, because wear and tear through erosion around the seating is more evenly distributed than in any other type of valve. A study of the other types shows that they all wear more at one point than another. Globe valves have reasonably short up and down movement of the disk, theoretically equal to a quarter of the seating diameter, and are not, therefore, big time wasters when being operated.

The disadvantage of the globe is, of course, the internal shape of the body; in providing an under and over flow, the diversion from the straight line creates a loss of pressure which, particularly on hydraulic installations, has to be carefully considered. It is doubtful if there would be any other type of valve for general purposes if it were not for this fact, and the reason for the introduction of the streamlined globe valve, with its reduced pressure loss, is to widen the applications of this most useful member of the valve family. The more head loss

(pressure loss) due to valves, bends, lengths of pipe, the bigger the pump to force the water round the system, and the larger the power consumption.

Plug Valve
Advantages of the plug valve are the straight through flow, with consequent minimum head loss, and the quick action requiring only 90° movement, i.e., quarter-turn.

The disadvantages of the basic plug cock are two fold. Firstly, its quick action can lead to water hammer on hydraulic installations when it is closed too rapidly, and secondly, it is difficult to combine tight shut-off with ease of operation.

The essential difference between the plug cock and the plug valve is that the latter incorporates features to reduce the friction between the plug and the body during operation, and to seal them against leakage. Probably the best-known method, which has been used for many years, is to introduce a 'lubricant', specially formulated for the purpose, between the plug and body. Such pressure lubricated plug valves are widely used for high pressure and high temperature applications.

Another method of rather more recent date is to interpose a sleeve, usually of PTFE, between the plug and the body to avoid metal to metal contact. In these sleeved plug valves it is usual for the sleeve to be retained in the body and for the plug to rotate within it. This method provides excellent shut-off and is particularly suited to applications where alloy materials are needed. The temperature limitation introduced by the use of PTFE in these valves must be recognized, however.

A derivative of the plug cock is the ball valve. Its advantages and disadvantages are similar to those of the sleeved plug valve, since it also usually employs seatings of PTFE. It tends to require even less torque to operate it than the sleeved plug valve, but its sealing security in severe applications tends to be less over a period of service.

An even further development of the plug valve is the butterfly valve – a compact, quick acting, and easy to operate valve which is well suited to flow regulation. A disadvantage is that to be drop tight the valve must again have a resilient seating which limits its use at high temperatures according to the elastomer used. Also, the disk has to be more substantial for higher pressures and because it is always in the flow path the head loss may be three times that of a gate valve.

GATE VALVE

The main advantage which can be claimed for the gate valve is that it has straight through flow when fully open and, therefore, minimum resistance to flow.

Its disadvantages are that it is the slowest acting of all valves because the gate has to be moved a distance greater than the bore of the valve, which also means that the valve is of necessity bulky in height. It is, however, relatively free from mechanical problems, and therefore remains the number one choice for hydraulic service.

DIAPHRAGM VALVE

This valve has two main advantages. No separate stem gland is necessary, as the diaphragm both shuts off the flow and acts as a gland. Also, the flexibility of the diaphragm provides a positive shut-off even on dirty fluids.

Its inherent disadvantage is that the diaphragm has to be made from an elastomeric material, usually either a natural or synthetic rubber, and this limits the application of the valve in respect to both temperature and pressure. Also, because of the stresses induced in the diaphragm, the valve has a shorter working life than other valve types.

From the analysis made so far the difficulty of imagining a single solution which will satisfy all requirements is easily seen and it seems that the problem of choice will exist for a long time to come.

Approaching this problem of valve selection in a logical manner, on the principles considered in the foregoing, we need first to know something about the flow resistance of the various types of valves.

Whatever conveys a liquid or a gas offers some resistance to flow. With a length of straight pipe, the resistance is caused by boundary friction between the internal surface of the pipe and the medium passing through it. When a valve is inserted into a pipeline it adds its own resistance to flow. With a plug or gate type of valve the flow is straight through and resistance is therefore low. In the case of a globe valve, where there are cavities and the flow can be deflected through as many as three right angles, the resistance is high.

Resistances to flow are found by experiment, and are usually expressed in equivalent lengths of straight pipe. This is a convenient way of expressing resistance because in this form it can be added straight into the general circuitary dimensions to obtain the total equivalent pipe lengths of an installation, and thence, by the use of standard tables, the total head loss or resistance characteristic determined.

Within quite narrow limits, a particular basic type of valve can be described as having a constant number of metres of straight pipe equivalent per millimetre of valve size. For a conventional globe valve this constant is $0 \cdot 4$ and therefore the resistance of a 25 mm globe valve is equal to that of $0 \cdot 4 \times 25 = 10$ m of 25 mm pipe while the resistance of a 50 mm globe valve is equal to $0 \cdot 4 \times 50 = 20$ m of 50 mm pipe, and so on.

CHART 3

The resistance constants of the various valve types are shown in simple diagrammatic form in Chart 3.

It can be seen that the rotating and sliding valves offer almost no resistance. The obturating types vary quite appreciably from the 'Y' type, which is really a globe valve with the headwork inclined at an angle to the pipe, at $0 \cdot 16$ m per mm, to the conventional globe type at $0 \cdot 4$ m per mm of size.

If it is decided that the obturating type is best suited for the purpose in mind, all that remains is the question of deciding from a resistance point of view whether the conventional globe valve, the streamlined globe valve, or the 'Y' globe valve should be chosen.

With the basic theory, the four basic valve types, their individual advantages and disadvantages and resistances to flow, there is now enough information to consider valve selection. The first requirement is to establish quite clearly the precise duty the particular valve is to perform. This can be done by providing answers to three easy questions:

(1) *Is the valve for use on a gas, including steam and air, or a liquid?*
The answer is important because when conveying liquids, head losses are far more important than when conveying gases.

For a liquid, a straight-through type of valve with minimum head loss should be used. With a gas, this is not so important.

(2) *Is the valve to be used often?*
The operator has to be considered before this question can be answered. If the valve is to be used frequently, the time taken in opening and closing the valve could be unacceptable.

For frequent use, therefore, the quick acting plug type should be chosen, with the obturating type as the second choice, and the sliding type avoided.

(3) *Is the valve only going to isolate the flow, that is be in either the on or the off position, or is it required to regulate the flow?*
As already seen, the obturating type of valve is the most suitable for throttling or regulation. All valves should, of course, unless otherwise qualified, be capable of isolation.

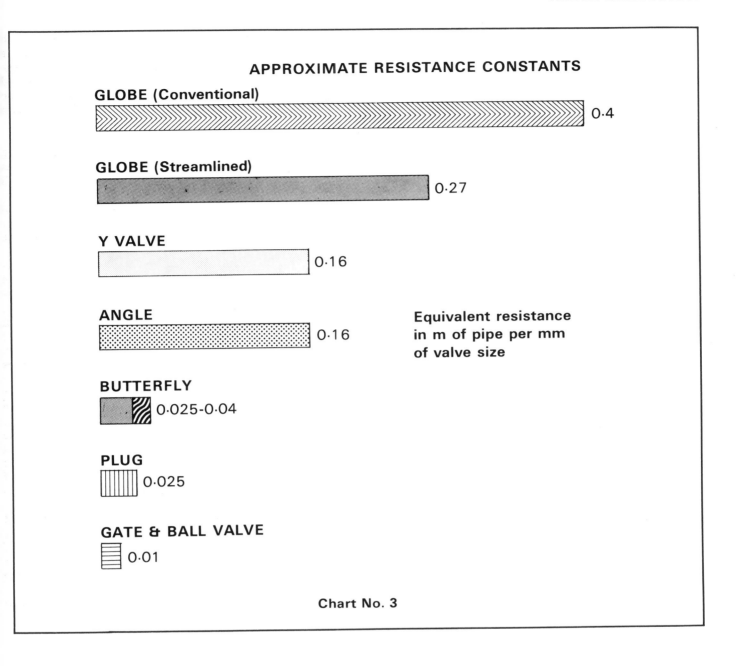

APPROXIMATE RESISTANCE CONSTANTS

GLOBE (Conventional) — 0·4

GLOBE (Streamlined) — 0·27

Y VALVE — 0·16

ANGLE — 0·16

Equivalent resistance in m of pipe per mm of valve size

BUTTERFLY — 0·025–0·04

PLUG — 0·025

GATE & BALL VALVE — 0·01

Chart No. 3

CHART 4

Obviously the answers given to these three questions can cause conflict in choosing the ideal valve, and accordingly it is necessary to compromise. Chart 4 is an attempt to give the best compromise solutions to the answers to the questions.

The chart is used in the order of answering the questions, connecting 'gas' or 'liquid' to 'frequently' or 'infrequently' and then dropping a perpendicular from either 'regulate' or 'isolate'. The point of intersection gives the best overall choice, while of the other two panels the one nearer to the point of intersection would be the second choice.

In the example shown on the chart, a valve for liquid (used frequently for isolation), a rotating type of valve would be chosen as this gives straight through flow and the quick action needed for frequent use. If, however, a rotating type to meet either pressure or temperature conditions cannot be obtained then the second choice would

be the obturating type because, for a valve to be fully opened and closed frequently, more emphasis would be given to the time required to operate the valve than to the straight through flow characteristic of the slow acting sliding type valve.

It is interesting to note that there are eight possible answers to the three questions and therefore eight different types of valving conditions. Obturating valves provide the answer to two of these, as do sliding valves. Rotating valves, however, fulfil four sets of conditions, which probably explains the increasing popularity of sleeved plug valves, ball valves, and butterfly valves.

Having decided from the diagram what basic type of valve would be most suitable, it is only necessary to qualify this choice according to the working pressure, temperature, and, in the case of obturating valves, the allowable pressure loss.

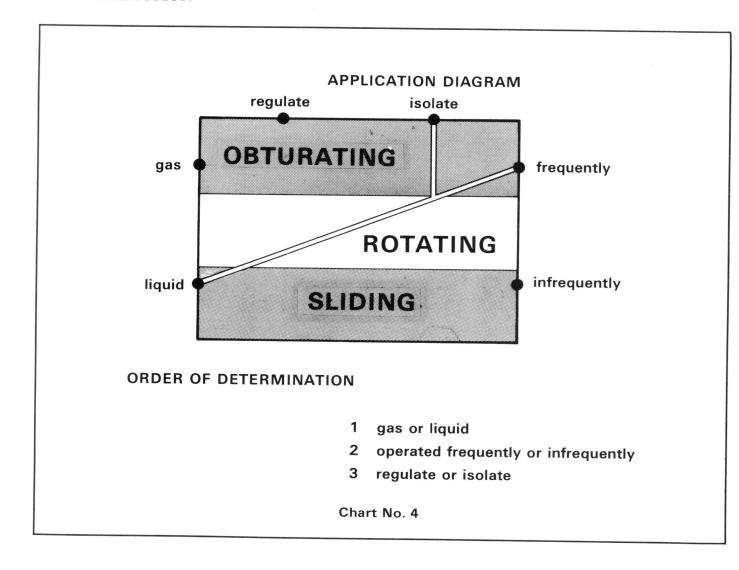

APPLICATION DIAGRAM

ORDER OF DETERMINATION

1   gas or liquid

2   operated frequently or infrequently

3   regulate or isolate

Chart No. 4

CHART 5

Obturating (Globe) Valves

These are available in all materials and all types of seating designs to fill every possible role. Therefore the only factor to consider is resistance to flow, the principal disadvantage of the globe valve. Decide, therefore, between conventional globe, streamlined globe, and 'Y' valve on the basis of flow resistance.

Rotating (Plug) Valves

The ordinary pattern is suitable for infrequent use on gas and water, while the lubricated type is preferable for the larger sizes and higher pressures. The sleeved plug valve and ball valve are more suitable for frequent use.

Sliding (Gate) Valves

This valve must be the number one choice for hydraulic applications where used infrequently, but operation should not be too infrequent. Gate valves benefit from being used occasionally, as this helps to prevent build-up of sediment on the seat faces which, when the valve is open, are exposed to the flowing medium.

Many seating variations are available. For example, PTFE inserts can be provided to give tighter shut-off on both gases and low surface tension fluids. PTFE is a synthetic material with a degree of softness which makes it suitable for a seating material and at the same time

has the temperature, chemical, and ageing resistance necessary for long service.

The butterfly valve can be considered as an alternative to the gate valve, particularly as size increases, since it requires much less headroom for installation. As seen earlier, the disk of a butterfly valve simply rotates within its own bore and is quicker in operation and more suited to regulation than the gate valve.

The parallel slide valve is best suited where stresses are caused by pipe expansion and contraction, for example on steam and high pressure hot water heating circuits.

The essential difference between the parallel slide valve and the wedge gate valve is that, as the names imply, the gate in the parallel slide valve has faces which are parallel and in opening slide across the parallel seatings in the body of the valve. The wedge gate valve, or sluice valve as it is sometimes called, has a tapered wedge which mates by a 'wedging action' with similarly tapered seats in the valve body.

When valves are designed it is with a particular use in mind. This article has simply reversed this procedure to show that the valves which will have the best chance of giving satisfaction in service can be chosen by considering their purpose first, and it is hoped that the Application Diagram and the Final Selection Chart will prove of value to all valve users.

FINAL SELECTION

obturating GLOBE valves

Decide between conventional globe, streamline globe and "Y" valve on basis of flow resistance.

---

rotating PLUG valves

Ordinary type for infrequent use on gas and water with lubricated type for larger sizes and higher pressures. Sleeved Plug and Ball valves for frequent use.

---

sliding GATE valves

No. 1 choice for hydraulic service with infrequent use.
Butterfly valve can be considered as alternative as size increases requiring less headroom and giving better regulation.

Chart No. 5

VALVE INSTALLATION AND MAINTENANCE

Valves which have been correctly selected in accordance with their service conditions should give years of trouble-free performance if correctly installed and regularly maintained.

Manufacturers' instructions regarding the installation and maintenance of particular valve types should always be consulted and the following information is intended only as a basic guide.

INSTALLATION

Wherever practicable, valves should be installed where there is adequate space available so that they can be conveniently operated and maintained.

Before installing a valve, check to ensure that size, pressure rating, materials of construction, end connections, etc. are suitable for the service conditions of the particular application.

Care must be taken to ensure that all dirt which may have accumulated in the valve during storage is removed before installation; maintain cleanliness during installation since the introduction of dirt can result in damage to the valve seats and operating mechanism. To facilitate cleanliness manufacturers usually fit covers on open ends.

To minimize the danger of abrasive particles damaging the seats, pipeline strainers should be fitted upstream of important valves.

Valves are precision units and as such their smooth operation will be impaired if they are subjected to line distortion stresses. Such strains must be avoided by ensuring that flanges and pipework are square and true and that pipes are properly supported to prevent the line buckling under the weight of the valve.

Particular care is necessary when welding valves into the line. Considerable distortion, resulting in line strains, may occur if valves are not welded into the line with care and the weld properly stress-relieved. Slag spatter should be avoided as its presence can be detrimental to the performance of the valve.

Pipework systems must always be subjected to testing prior to becoming fully operational. Although valves are tested prior to despatch from the manufacturer it is probable that some adjustment will be necessary, for instance to the gland packing, when the valve is on stream. Line testing will also prove the quality of flange bolting, welding, etc.

MAINTENANCE

As with all mechanical devices, regular maintenance is the most efficient means of ensuring continued operational efficiency.

It is recommended that in every case the procedure followed is that given in the manufacturer's maintenance instructions.

Regular scheduled inspection of all valves is essential and especially of those valves which are operated only occasionally, such as isolation and emergency valves.

Continuing efficient operation of strainers must also be checked and maintenance carried out at the same time as that on valves.

Valves having gland packings should be inspected frequently to check that the pressure seal is being maintained. At the first sign of leakage, the gland should be adjusted or, if necessary, the packing replaced.

Examine valve disks and seatings to ascertain the extent of any damage or service wear; either recondition on site or return to the manufacturer.

Cover and flange gaskets should be inspected and replaced where necessary.

SPARES

An adequate and readily available stock of spares for valves is an essential requirement at all plants.

Manufacturers will be pleased to advise on suitable schedules of spare parts for particular valves and circumstances.

Spare parts obviously need to be protected from exposure and it is usual to coat them with anti-corrosive solution. It is desirable that spares are packed and labelled so as to be readily identifiable on site.

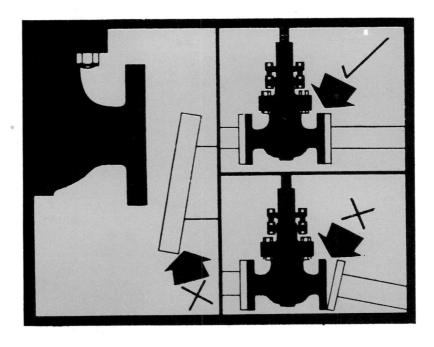

Figs. 1 and 2. The accompanying diagrams
have been reproduced, with permission, from
a series appearing in a well known
manufacturer's Installation and Maintenance
Manual

RECONDITIONING OF VALVES

Reconditioning of valves has become a significant part of the supply of valves to industry. Given reasonable care and regular maintenance, most industrial valves will give long and satisfactory service but in the course of time the operational efficiency of a valve can fall to some unacceptable level. Consideration may then well be given to having the valve reconditioned as an alternative to the purchase of a completely new valve.

In many cases, especially where a large valve or a valve manufactured in expensive materials is under considera-tion, reconditioning can offer substantial cost savings compared with the expenditure required for a new valve. Also, a replacement valve may be available only on an extended delivery time and reconditioning can often pro-vide a worthwhile saving in this direction.

It is always advisable to seek the advice of the original manufacturer when undertaking reconditioning involving replacement of parts – particularly when more specialized materials form part of the product.

QUALITY CONTROL

Quality control is concerned with three main aspects – quality of materials, quality of components, and quality of the finished valve.

Quality control should not be confused with quality assurance. The latter is concerned with ensuring that the quality control and the systems used are effective in maintaining the standards specified.

Basic quality control requirements are indicated, but not necessarily specified, by the standard specification to which the valve is made, or covering valves for a similar application if the valve is such that it is not covered by a particular standard specification. For many applications involving valves in regular production, careful control of material supplies, visual and dimensional inspection of components, and pressure testing of the final valve at ambient temperature will be sufficient.

The extent of quality control to be applied in the manufacture of a particular valve should be determined primarily in relation to its application. Manufacturers generally apply quality control appropriate to the types of valves which they produce and the normal uses of such valves. They will readily advise what are their standard control procedures for particular valves and these should prove more than adequate for most purposes. Obviously, this will vary according to the type of valve, being more exhaustive for a high-pressure steam valve than for a bronze radiator valve.

Where valves are of special design, made of special materials as compared with normal production, or for critical services, then more exhaustive quality control may be considered desirable. Whenever this is thought to be the case, it should be discussed in detail with the prospective supplier.

Materials

It is necessary to ensure that materials used are correct as specified and that they are in a suitable condition for the purpose intended. This latter will involve general inspection and may or may not include nondestructive testing procedures appropriate to the form, such as ultrasonic testing, radiography, and crack detection. The scope and extent of NDT should be determined in relation to the severity and hazard of the end use of the valve.

For items in regular volume production, some form of sampling is often used as material quality control and again the extent of testing is determined by the intended purpose of the valve.

Components

When material quality control has been applied, quality control on components is basically one of inspection against the design requirements. Again, the application of total inspection or the use of sampling must be determined by circumstances.

Finished Valve

Here, quality control is concerned with satisfactory operation and function of the valve in accordance with the specification to which it is made. This will involve both inspection and testing, usually including pressure testing both of the pressure envelope and of the seating, where valves have a primarily shut-off function.

Many valve specifications, including British Standards, require that there shall be no visible leakage through the valve seat under specified test conditions. In practice it is appreciated that a valve conforming to this requirement on initial test may not necessarily produce the same result on later tests carried out under different conditions.

Some valve product standards have acknowledged the impracticability of zero seat leakage concepts for certain industrial valves and in these maximum leakage rates are specified. Specified permissible leakage should be based on process considerations such as hazards resulting from leakage after emergency shut-off, etc. In many industrial applications, some slight degree of leakage is acceptable and permissible.

All aspects of quality control, inspection, and testing should be carefully weighed against the application and hazard associated with the service of the valve. Caution should be exercised to avoid over-specification, which can result in an unnecessarily expensive product and which may reduce sources of availability and extend delivery times.

Any requirements for quality control (and for material and test certificates if they are thought to be necessary) should be discussed and agreed with the supplier at the time of an order, since changes made after work has commenced may lead to increased costs due not least to wasted work, and consequently to extended delivery times.

NOISE AND INDUSTRIAL VALVES

INTRODUCTION

Noise, which can be defined as 'unwanted sound', is increasingly being regarded as a pollutant which detracts from the quality of life. Exposure to excessive noise, in addition to being extremely unpleasant, can result in permanent damage to hearing. Relatively low levels of noise at night may be unacceptable and yet pass unnoticed in the general sound level of daytime activity.

Noise Measurement

Most noise measurements are given in decibels (dB). The reason for using a relative instead of an absolute scale is because of the magnitude of the pressure range involved. The smallest sound pressure to which a normal person's hearing responds is $0 \cdot 0002$ μbar. At levels above 200 μbar sound is felt as well as heard, while above 2000 μbar the pressure levels are unbearable. It is more convenient to measure the actual sound pressure, compare it with the reference level ($0 \cdot 0002$ μbar), and then express the result in decibels. Thus the sound pressure level (SPL) can be given by a scale between 0 and 120 dB. In mathematical terms:

$$\text{SPL} = 20 \log \frac{P}{P_0},$$

where P = measured pressure in μbar, and
$\quad P_0 = 0 \cdot 0002$ μbar.

When two noise producing sources are in close proximity the combined noise is calculated as follows:

$$\text{Total SPL} = 10 \log \left(\text{antilog} \frac{\text{SPL}_1}{10} + \text{antilog} \frac{\text{SPL}_2}{10} \right),$$

where SPL_1 is the noise emitted from source 1, and
$\quad \text{SPL}_2$ is the noise emitted from source 2.

An increase of 3 dB represents a doubling of the sound intensity.

Sound is attenuated as the distance from the source to the observer is increased. Assuming linear radiation, the noise at a distance can be calculated using the following formula:

$$\text{SPL}_{\text{dist.}} = \text{SPL} - 10 \log (\text{distance in metres})$$

where SPL is the noise emitted from the source (at 1 metre).

Sound pressure level is normally measured in decibels using instruments with the A-weighted scale, i.e. giving readings in dBA. The A-scale is used because it is most closely related to the response of the human ear. Alternatively, an octave band analysis can be made over the whole frequency range and then converted to an overall dBA rating.

When making noise measurements from a valve, it is usual to take readings at points situated 45° from the pipeline (both upstream and downstream) and at a radius of 1 metre away from the valve body.

Typical Noise Levels

Some idea of relative noise levels is given by the following examples:

130 dB	jet engine at 30 m;
100 dB	inside underground train;
80 dB	inside small car;
60 dB	normal conversation at 1 m;
40 dB	public library;
20 dB	quiet country lane.

VALVE NOISE

Under certain conditions of service industrial valves can generate noise. In many cases it is possible to predict whether a valve will be noisy under certain conditions of service. In general it is those valves which operate at various openings that give the most trouble. Valves having on–off duties give less difficulty in this respect. The control valve is an example of a valve type having variable opening in service.

Sources of Valve Noise

There are three main types of valve noise – all are generated by the passage of the line fluid through the interior of the valve. Each can be avoided or alleviated but the treatment may be expensive.

Mechanical vibration is induced by the passage of the fluid through the valve. Slackness in bearings or guides can give rise to oscillation of the internals. In severe cases the trim may go into resonance followed by rapid fatigue failure. Possible remedies for this type of problem are:

(1) Better guiding or closer running fits in existing bearings.
(2) Change of mass or stiffening up of the plug.
(3) Reversal of the flow direction through the valve.
(4) Change of pressure drop across the valve.

Cavitation noise is caused by the collapse of vapour bubbles in a liquid due to pressure recovery downstream of the valve orifice. Although the noise level produced is not high the physical damage to internals can be extensive and rapid. Cavitation can be alleviated by reducing the pressure drop across the valve, by fitting valves which have very little pressure recovery, or by using specially designed equipment with multiple velocity head loss trim.

Aerodynamic noise is produced by a gas accelerating to supersonic velocity at critical or higher pressure drops through the trim. The resulting shock waves and general turbulence at the fluid boundary generate sound which travels downstream. Under certain conditions it may also be propagated upstream of a valve. This noise can be worst of all, ranging from a whistling sound to a heavy roar, e.g., when a boiler safety valve is discharging.

Apart from health considerations or infringement of legal requirements, other undesirable effects may be promoted by vibrations induced in the structure. Cases have been reported of damage to valve internals, fracture of welded joints in pipelines and flanges, breakage of instrument impulse connections, and loosening of nuts and bolts on high pressure connections. Even though a silencing device may be fitted downstream of the source it does nothing to damp the vibrations or their damaging effects. As always, prevention is better than cure so a check on the flowing conditions is usually a worthwhile exercise in a noise sensitive situation.

NOISE REDUCTION TECHNIQUES

In general there are two cases for consideration:
  (1) Before final valve selection, examination of the flow data suggests the probability of a noise problem.
  (2) After commissioning, there is a noise problem and a means of reducing or eliminating it is desired.

Case 1
Calculation methods are available from various valve manufacturers who have researched the subject. Some are mathematical and others are based on the results of practical tests; all give a measure of the probable sound level. Having established that an unacceptable level is likely, some means of reducing it have to be considered.

Special equipment is available to cater for many conditions of service. By fitting special valves or special internals to standard valves it may be possible to arrive at an acceptable solution within the valve itself. If this cannot be done then the provision of one or more fixed area devices downstream of the valve can be considered. Fitting heavy wall piping downstream of the valve will alleviate some of the sound; similarly, a silencer will reduce the level; but neither will do anything to modify the vibration. Reducing the pressure drop across the valve will help either by increasing the downstream pressure or by reducing the upstream pressure. Increasing both the upstream and downstream pressures may be tried, providing the P_1/P_2 ratio comes down in value.

The problems involved in rectifying an already noisy installation can be considerable. Possible remedies are:

Fit a second valve – two valves in series will reduce the pressure drop across each to what may be a more acceptable level.

Fit a fixed area restriction downstream of the valve.

Put a silencing unit on the downstream piping.

Put the valve in a pit and bury the downstream piping for as far as possible.

Install special valves to control noise emission.

Use heavy wall piping downstream and insulate with acoustic lagging.

Change the pressure drop across the valve to a lower P_1/P_2.

One or more of the above may give the desired result but the cure can be expensive. It is always better to anticipate a noisy situation and cater for it than have to try to remove the difficulty later on.

NOISE LEGISLATION

In the United Kingdom the Health and Safety at Work etc. Act 1974 refers in general terms to excessive noise in the working environment. More specifically, the *Code of Practice for Reducing the Exposure of Employed Persons to Noise* (issued by the Department of Employment) lays down the rules, although it does not have the force of law. An analysis of the rules is not given since the Code of Practice is subject to change. Furthermore, other countries have their own legislation on the subject. It is important therefore, to check on the latest regulations pertaining in the country in which the valve is to be installed.

FLOW DATA

PRESSURE DROP THROUGH A VALVE

Liquid Flow

Head loss through a valve can be expressed by the equation

$$\Delta h = \frac{Kv^2}{2g}$$

where Δh = head loss in metres of liquid,
K = resistance coefficient,
v = mean flow velocity in m/s,
g = acceleration of gravity, $9 \cdot 81$ m/s².

By substitution of proper units, the above equation may be written to express the pressure drop equivalent to Δh:

$$\Delta P = 0 \cdot 000\,005\ Kv^2 \rho$$

where ΔP = pressure drop in bar
ρ = density of liquid in kg/m³.
or, in terms of flow rate:

$$\Delta P = 1 \cdot 389\ \frac{KQ^2 G}{a^2}$$

where Q = flow rate in l/min,
G = specific gravity of the liquid,
a = nominal flow area of valve in mm².

Gaseous Flow

In the case of compressible fluids some modifications may be required to the above formula and users are advised to consult the valve manufacturer.

EQUIVALENT LENGTH

The length of straight pipe which would offer the same amount of resistance as the valve can be computed from the equation

$$L = \frac{dK}{1000f} \qquad \text{see Note (1)}$$

where L = equivalent length of pipe in m,
d = diameter of pipe in mm,
f = friction factor for pipe.

This can be a convenient way of expressing valve resistances as the equivalent lengths can be added to that of the general piping circuit to obtain the total equivalent pipe length and the total head loss found by the use of established tables.

The chart of resistance constants provided in the section on valve selection is based on the equivalent length method of expressing valve resistances.

Notes: (1) The friction factor f used in the equation for equivalent length is the Darcy friction factor. In some literature on the subject a friction factor one-fourth the value of the Darcy friction factor is used and the equation then reads $L = dK/4000f$.
(2) The resistance coefficient K and friction factor f must be obtained from the valve manufacturer.

FLOW COEFFICIENTS

It is often convenient, particularly with control valves, to be able to express the relationship between pressure drop and flow rate through a valve by a flow coefficient. In many parts of the world, including the UK and the USA, the flow coefficient in most general use is C_v, while in parts of Europe the coefficient K_v is often used. Another coefficient, A_v, is also available, although this is directly associated with SI units.

Each type and size of valve has a particular flow coefficient and the general formulae which follow indicate how a knowledge of this can be used to establish the pressure drop across a valve for a given flow rate or, alternatively, to determine the flow rate through a valve which will generate a given pressure drop.

Flow coefficient values are determined by testing and may not be valid for all conditions of flow. Hence, users requiring information on the flow characteristics of a valve are advised always to consult with the valve manufacturer.

Flow Coefficient C$_v$

This may be defined as:
    The rate of flow of water in US gallons per minute, at 60 °F (15·6 °C), that will generate a pressure drop of one pound-force per square inch across the valve.

LIQUID FLOW

The basic formula is:

$$C_v = \frac{Q\sqrt{G}}{\sqrt{(\Delta P)}}$$

where Q = flow rate in US gal/min,
    G = specific gravity of liquid (water = 1),
    ΔP = pressure drop across valve in lbf/in^2.
(This assumes the flow to be neither viscous, cavitating, nor flashing.)

GASEOUS FLOW

Volume Flow

$$C_v = \frac{Q}{\sqrt{\dfrac{289}{GT}}\;\; C_1 C_2 P_1 \sin\left[\dfrac{3417}{C_1 C_2}\;\sqrt{\left(\dfrac{\Delta P}{P_1}\right)}\right]\text{deg.}}$$

Weight Flow

$$C_v = \frac{4\cdot32 W\sqrt{T}}{C_1 C_2 P_1 \sqrt{M}\sin\left[\dfrac{3417}{C_1 C_2}\;\sqrt{\left(\dfrac{\Delta P}{P_1}\right)}\right]\text{deg.}}$$

where Q = gas flow rate, s.c.f.h.,
    G = specific gravity of gas (air = 1),
    T = temperature of gas in K (= °C + 273),
    C_1 = type-of-valve constant,
    C_2 = correction factor for ratio of specific heats,
    ΔP = pressure drop across valve in lbf/in^2,
    P_1 = valve inlet pressure in lbf/in^2 abs.,
    deg. = angular degrees,
    M = molecular weight of gas,
    W = gas flow rate in lb/h.

STEAM FLOW

$$C_v = \frac{W\sqrt{V_1}}{1\cdot06 C_1\sqrt{P_1}\times\sin\left[\dfrac{3417}{C_1}\;\sqrt{\left(\dfrac{\Delta P}{P_1}\right)}\right]\text{deg.}}$$

where W = steam flow rate in lb/h,
    V_1 = specific volume at upstream conditions in ft^3/lb,
    C_1 = type-of-valve constant,
    P_1 = valve inlet pressure in lbf/in^2 abs.,
    ΔP = pressure drop across valve in lbf/in^2,
    deg. = angular degrees.
(If the bracketed quantity equals or exceeds 90° then the flow is critical and sin[   ] = 1, irrespective of the actual calculated value.)

In the above formulae for gas and steam flow, the value of constant C_1 must be obtained from the valve manufacturer.
    Notes: (1) The formulae for C_v are shown in terms of British units as, at the time of printing, no agreement had been reached regarding the specific SI units to be used for this coefficient.
    (2) The formulae for gas flow and steam flow presented here are included in the current draft IEC Standards on control valve sizing. Other forms of these formulae are also available.

Flow Coefficient K$_v$

This may be defined as:
    The rate of flow of water in cubic metres per hour that will generate a pressure drop of one bar across the valve.
    The basic equation is:

$$K_v = \frac{Q\sqrt{G}}{\sqrt{(\Delta P)}}$$

where Q = flow rate in m^3/h,
    G = specific gravity of liquid (water = 1),
    ΔP = pressure drop across valve in bar.
(This assumes the flow to be neither viscous, cavitating, nor flashing.)

Flow Coefficient A$_v$

As already stated, this coefficient is associated with SI units. Methods of evaluating control valve capacity using flow coefficient A_v are described in a British Standard prepared under the authority of the Instrument Industry Standards Committee and deal with incompressible fluids, gases, and vapours. All units used are in accordance with the International System (SI).

LIQUID FLOW

The basic formula is:

$$A_v = \frac{Q\sqrt{\rho}}{\sqrt{(\Delta P)}}$$

where Q = flow rate in m^3/s,
    ΔP = pressure loss in pascals,
    ρ = density of liquid in kg/m^3.

GASEOUS FLOW

The recommended working formula is:

$$A_v = \frac{W}{1 \cdot 67 \times 10^{-2} \times C_1 C_2 Y \sqrt{(\rho_1 P_1)}}$$

where W = flow rate in kg/s,
C_1 = gas flow factor (type-of-valve constant),
C_2 = factor for variation in specific heats,
ρ_1 = density of fluid at inlet in kg/m^3,
P_1 = pressure of fluid at inlet in pascals,
Y = subcritical flow factor, being given by:

$$Y = \sin \left[\frac{3417}{C_1 C_2} \sqrt{\frac{(\Delta P)}{P_1}} \right] \text{ degrees}$$

where ΔP = pressure drop in pascals.

If the expression in the brackets equals or exceeds 90°, then the flow is critical and, in this case, Y is unity.

Notes: (1) A_v has the dimensions of area (m^2) but is not the area of the valve orifice.

(2) For full information concerning flow coefficient A_v, reference should be made to the relevant British Standard and to the IEC Standards in the series 534.

CONVERSION TABLES

Where relevant in the following tables of conversion factors the SI unit or multiple thereof recommended by the British Valve Manufacturers' Association for use in the valve industry is shown in the left hand column.

The degree of rounding of the conversion factors, which are generally based on BS 350, has been adjusted to an extent considered to be of value to a practical engineer.

Length

millimetre mm	centimetre cm	metre m	inch in	foot ft	yard yd
1	0·1	0·001	0·0394	0·0033	0·0011
10	1	0·01	0·3937	0·0328	0·0109
1000	100	1	39·3701	3·2808	1·0936
25·4	2·54	0·0254	1	0·0833	0·0278
304·8	30·48	0·3048	12	1	0·3333
914·4	91·44	0·9144	36	3	1

1 kilometre = 1000 metres = 0·621 37 miles
1 mile = 1609·34 metres = 1·609 34 kilometres

Area

square millimetre mm^2	square centimetre cm^2	square metre m^2	square inch in^2	square foot ft^2	square yard yd^2
1	0·01	10^{-6}	$1·55 \times 10^{-3}$	$1·076 \times 10^{-5}$	$1·196 \times 10^{-6}$
100	1	10^{-4}	0·155	$1·076 \times 10^{-3}$	$1·196 \times 10^{-4}$
10^6	10 000	1	1550	10·764	1·196
645·16	6·4516	$6·452 \times 10^{-4}$	1	$6·944 \times 10^{-3}$	$7·716 \times 10^{-4}$
92 903	929·03	0·093	144	1	0·111
836 127	8361·27	0·836	1296	9	1

Volume

cubic millimetre mm^3	cubic centimetre cm^3	cubic metre m^3	cubic inch in^3	cubic foot ft^3	cubic yard yd^3
1	0·001	10^{-9}	$6·1 \times 10^{-5}$	$3·531 \times 10^{-8}$	$1·308 \times 10^{-9}$
1000	1	10^{-6}	0·061	$3·531 \times 10^{-5}$	$1·308 \times 10^{-6}$
10^9	10^6	1	61 024	35·31	1·308
16 387	16·39	$1·639 \times 10^{-5}$	1	$5·787 \times 10^{-4}$	$2·143 \times 10^{-5}$
$2·832 \times 10^7$	$2·832 \times 10^4$	0·0283	1728	1	0·0370
$7·646 \times 10^8$	$7·646 \times 10^5$	0·7646	46 656	27	1

Volume (Capacity)

litre l	cubic metre m^3	millilitre ml	UK gallon UK gal	US gallon US gal	cubic foot ft^3
1	0·001	1000	0·22	0·2642	0·0353
1000	1	10^6	220	264·2	35·3147
0·001	10^{-6}	1	$2·2 \times 10^{-4}$	$2·642 \times 10^{-4}$	$3·53 \times 10^{-5}$
4·546	0·004 55	4546	1	1·201	0·1605
3·785	0·003 78	3785	0·8327	1	0·1337
28·317	0·0283	28 317	6·2288	7·4805	1

1 US barrel = 42 US gallons (petroleum measure)
1 litre = 10^6 mm^3 = 10^3 cm^3 or 1 cubic decimetre (1 dm^3)
1 litre = 1·76 UK pints

Velocity

metre per second m/s	foot per second ft/s	metre per minute m/min	foot per minute ft/min	kilometre per hour km/h	mile per hour mile/h
1	3·281	60	196·85	3·6	2·2369
0·305	1	18·288	60	1·0973	0·6818
0·017	0·055	1	3·281	0·06	0·0373
0·005	0·017	0·305	1	0·0183	0·0114
0·278	0·911	16·667	54·68	1	0·6214
0·447	1·467	26·822	88	1·6093	1

Mass

kilogram kg	pound lb	hundredweight cwt	tonne t	UK ton ton	US ton sh tn
1	2·205	0·0197	0·001	$9·84 \times 10^{-4}$	0·0011
0·454	1	0·0089	$4·54 \times 10^{-4}$	$4·46 \times 10^{-4}$	$5·0 \times 10^{-4}$
50·802	112	1	0·0508	0·05	0·056
1000	2204·6	19·684	1	0·9842	1·1023
1016	2240	20	1·0161	1	1·12
907·2	2000	17·857	0·9072	0·8929	1

Mass Flow Rate

kilogram per second kg/s	pound per second lb/s	kilogram per hour kg/h	pound per hour lb/h	UK ton/hour ton/h	tonne/hour t/h
1	2·205	3600	7936·64	3·5431	3·6
0·454	1	1633	3600	1·607	1·633
$2·78 \times 10^{-4}$	$6·12 \times 10^{-4}$	1	2·205	$9·84 \times 10^{-4}$	0·001
$1·26 \times 10^{-4}$	$2·78 \times 10^{-4}$	0·454	1	$4·46 \times 10^{-4}$	$4·54 \times 10^{-4}$
0·282	0·622	1016	2240	1	1·016
0·278	0·612	1000	2204·6	0·9842	1

Volumetric Rate of Flow

litre per second l/s	litre per minute l/min	cubic metre per hour m³/h	cubic foot per hour ft³/h	cubic foot per minute ft³/min	UK gallon per minute UK gal/min	US gallon per minute US gal/min	US barrel per day US barrel
1	60	3·6	127·133	2·1189	13·2	15·85	543·439
0·017	1	0·06	2·1189	0·0353	0·22	0·264	9·057
0·278	16·667	1	35·3147	0·5886	3·666	4·403	150·955
0·008	0·472	0·0283	1	0·0167	0·104	0·125	4·275
0·472	28·317	1·6990	60	1	6·229	7·480	256·475
0·076	4·546	0·2728	9·6326	0·1605	1	1·201	41·175
0·063	3·785	0·2271	8·0209	0·1337	0·833	1	34·286
0·002	0·110	0·0066	0·2339	0·0039	0·024	0·029	1

Force

newton N	kilonewton kN	kilogram-force* kgf	pound-force lbf
1	0·001	0·102	0·225
1000	1	101·97	224·81
9·807	0·0098	1	2·205
4·448	0·0044	0·454	1

* The kilogram-force is sometimes called the kilopond (kp)

Moment of Force

newton metre N m	kilonewton metre kN m	kilogram-force metre kgf m	pound-force inch lbf in	pound-force foot lbf ft
1	0·001	0·102	8·85	0·738
1000	**1**	101·972	8851	737·6
9·807	0·0098	**1**	86·8	7·233
0·113	$1·13 \times 10^{-4}$	0·0115	**1**	0·083
1·356	0·0014	0·138	12	**1**

Energy, Work, Heat

joule J	kilojoule kJ	megajoule MJ	foot pound-force ft lbf	British thermal unit Btu	therm	kilowatt hour kW h
1	0·001	10^{-6}	0·738	$9·48 \times 10^{-4}$	$9·48 \times 10^{-9}$	$2·78 \times 10^{-7}$
1000	**1**	0·001	737·56	0·9478	$9·48 \times 10^{-6}$	$2·78 \times 10^{-4}$
10^6	1000	**1**	737 562	947·82	$9·48 \times 10^{-3}$	0·2778
1·356	0·0014	$1·36 \times 10^{-6}$	**1**	0·0013	$1·28 \times 10^{-8}$	$3·77 \times 10^{-7}$
1055·1	1·0551	0·0010	778·17	**1**	10^{-5}	$2·931 \times 10^{-4}$
$1·0551 \times 10^8$	105 510	105 51	$7·78 \times 10^7$	100 000	**1**	29·307
$3·6 \times 10^6$	3600	3·6	$2·65 \times 10^6$	3412·1	0·03412	**1**

1 joule = 1 newton metre

Power

watt W	kilogram-force metre per second kgf m/s	metric horsepower	foot pound-force per second ft lbf/s	horsepower hp
1	0·102	0·00136	0·738	0·00134
9·806	**1**	0·0133	7·233	0·0131
735·5	75	**1**	542·476	0·9863
1·356	0·138	0·00184	**1**	0·00182
745·70	76·04	1·0139	550·0	**1**

1 watt = 1 joule per sec = 1 newton metre per second

Heat Flow Rate

watt W	calorie per second cal/s	kilocalorie per hour kcal/h	British thermal unit per hour Btu/h
1	0·239	0·86	3·412
4·187	**1**	3·6	14·286
1·163	0·278	**1**	3·968
0·293	0·07	0·252	**1**

1 watt = 1 joule per sec = 1 newton metre per second

Pressure and Liquid Head

bar (10⁵ N/m²) bar	millibar (10² N/m²) mbar	pascal (1 N/m²) Pa	kilogram-force per square centimetre kgf/cm²	pound-force per square inch lbf/in²	foot of water ft H₂O	metre of water m H₂O	millimetre of mercury mm Hg	inch of mercury in Hg
1	1000	10⁵	1·02	14·5	33·455	10·2	750·1	29·53
0·001	1	100	1·02 × 10⁻³	0·0145	0·033	0·0102	0·75	0·029
10⁻⁵	0·01	1	1·02 × 10⁻⁵	1·45 × 10⁻⁴	3·3 × 10⁻⁴	1·02 × 10⁻⁴	0·0075	2·95 × 10⁻⁴
0·981	980·7	98 067	1	14·22	32·808	10·0	735·6	28·96
0·069	68·95	6895	0·0703	1	2·307	0·703	51·71	2·036
0·03	29·89	2989	0·0305	0·433	1	0·305	22·42	0·883
0·098	98·07	9807	0·1	1·42	3·28	1	73·55	2·896
0·0013	1·333	133·3	0·0014	0·019	0·045	0·014	1	0·039
0·0339	33·86	3386	0·0345	0·491	1·333	0·345	25·4	1

One millimetre head of mercury (1 mmHg) is also known by the name 'torr'.
The international standard atmosphere (1 atm) = 1·013 25 bar = 14·6959 lbf/in².
The technical (metric) atmosphere (1 at) = 0·980 66 bar = 14·2233 lbf/in².
The conventional reference conditions known as 'standard temperature and pressure' (s.t.p.) are: 1·013 25 bar (14·6959 lbf/in²) at 0°C.
The standard reference conditions (s.t.) for gas, as defined by the International Gas Union, are: 1·013 25 bar at 15 °C and dry. These conditions may also be referred to as 'Metric Standard Conditions' (MSC).

Stress

megapascal (10⁶ N/m²) MPa	pound-force per square inch lbf/in²	UK ton-force per square inch UK tonf/in²	kilogram-force per square centimetre kgf/cm²	kilogram-force per square millimetre kgf/mm²
1	145	0·0647	10·197	0·102
0·0069	1	4·464 × 10⁻⁴	0·0703	7·031 × 10⁻⁴
15·44	2240	1	157·5	1·575
0·0981	14·22	0·0063	1	0·01
9·807	1422	0·6350	100	1

A 'pascal' is the special name for one newton per square metre (1 Pa = 1 N/m²).
One megapascal is equal to one newton per square millimetre (1 MPa = 1 N/mm²).

Specific Enthalpy (and Specific Energy)

kilojoule per kilogram kJ/kg	kilocalorie per kilogram kcal/kg	British thermal unit per pound Btu/lb	kilogram-force metre per kilogram kgf m/kg	foot pound-force per pound ft lbf/lb	kilowatt hour per kilogram kW h/kg
1	0·239	0·4299	101·97	334·55	2·78 × 10⁻⁴
4·187	1	1·8	426·93	1400·7	11·63 × 10⁻⁴
2·326	0·555	1	237·19	778·17	6·46 × 10⁻⁴
0·009 81	0·002 34	0·004 22	1	3·28	2·72 × 10⁻⁶
0·002 99	7·139 × 10⁻⁴	0·001 28	0·305	1	8·30 × 10⁻⁷
3600	859·845	1547·72	3·671 × 10⁵	1·204 × 10⁶	1

Specific Volume

cubic centimetre per gram cm³/g	cubic metre per kilogram m³/kg	cubic inch per pound in³/lb	cubic foot per pound ft³/lb	cubic foot per UK ton ft³/ton	UK gallon per pound UKgal/lb	US gallon per pound USgal/lb
1	0·001	27·68	0·016	35·88	0·0998	0·1198
1000	**1**	27 680	16·018	35 881	99·776	119·83
0·036	$3·613 \times 10^{-5}$	**1**	$5·787 \times 10^{-4}$	1·296	0·0036	0·0043
62·428	0·062	1728	**1**	2240	6·229	7·481
0·028	$2·787 \times 10^{-5}$	0·771	$4·464 \times 10^{-4}$	**1**	0·0028	0·0033
10·022	0·010	277·4	0·160	359·6	**1**	1·201
8·345	0·0083	231·0	0·134	299·4	0·833	**1**

1 cm³/g = 1 ml/g = 1 dm³/kg = 1 litre/kg = 1 m³/tonne
1 m³/kg = 1 dm³/g

Density

gram per cubic centimetre g/cm³	kilogram per cubic metre kg/m³	pound per cubic inch lb/in³	pound per cubic foot lb/ft³	UK ton per cubic yard ton/yd³	pound per UK gallon lb/UKgal	pound per US gallon lb/USgal
1	1000	0·036	62·428	0·752	10·022	8·345
0·001	**1**	$3·613 \times 10^{-5}$	0·062	$7·525 \times 10^{-4}$	0·010	0·0083
27·68	27 680	**1**	1728	20·829	277·42	231
0·016	16·018	$5·787 \times 10^{-4}$	**1**	0·012	0·160	0·134
1·329	1328·9	0·048	82·963	**1**	13·319	11·090
0·0998	99·78	$3·605 \times 10^{-3}$	6·229	0·075	**1**	0·833
0·1198	119·83	$4·329 \times 10^{-3}$	7·480	0·090	1·201	**1**

1 g/cm³ = 1 g/ml = 1 kg/dm³ = 1 kg/litre = 1 tonne/m³
1 kg/m³ = 1 g/dm³

MASS PER UNIT LENGTH

1 kg/m = 0·6720 lb/ft
1 lb/ft = 1·488 kg/m

MASS PER UNIT AREA

1 kg/m² = 0·2048 lb/ft²
1 lb/ft² = 4·882 kg/m²

CALORIFIC VALUE (VOLUME BASIS)

1 MJ/m³ = 26·84 Btu/ft³
1 Btu/ft³ = 0·037 26 MJ/m³ or
          37·26 kJ/m³

THERMAL CONDUCTIVITY

1 W/(m °C) = 6·934 Btu in/(ft² h °F)
1 Btu in/ft² h °F = 0·1442 W/(m °C)
(i.e., 0·1442 W m/(m² °C))

Temperature Conversions

−459·4 to 0

°C	†	°F
−273	459·4	
−268	−450	
−262	−440	
−257	−430	
−251	−420	
−246	−410	
−240	−400	
−234	−390	
−229	−380	
−223	−370	
−218	−360	
−212	−350	
−207	−340	
−201	−330	
−196	−320	
−190	−310	
−184	−300	
−179	−290	
−173	−280	
−169	−273	−459·4
−168	−270	−454
−162	−260	−436
−157	−250	−418
−151	−240	−400
−146	−230	−382
−140	−220	−364
−134	−210	−346
−129	−200	−328
−123	−190	−310
−118	−180	−292
−112	−170	−274
−107	−160	−256
−101	−150	−238
−96	−140	−220
−90	−130	−202
−84	−120	−184
−79	−110	−166
−73	−100	−148
−68	−90	−130
−62	−80	−112
−57	−70	−94
−51	−60	−76
−46	−50	−58
−40	−40	−40
−34	−30	−22
−29	−20	−4
−23	−10	14
−17·8	0	32

0 to 49

°C	†	°F
−17·8	0	32
−17·2	1	33·8
−16·7	2	35·6
−16·1	3	37·4
−15·6	4	39·2
−15·0	5	41·0
−14·4	6	42·8
−13·9	7	44·6
−13·3	8	46·4
−12·8	9	48·2
−12·2	10	50·0
−11·7	11	51·8
−11.1	12	53·6
−10·6	13	55·4
−10·0	14	57·2
−9·4	15	59·0
−8·9	16	60·8
−8·3	17	62·6
−7·8	18	64·4
−7·2	19	66·2
−6·7	20	68·0
−6·1	21	69·8
−5·6	22	71·6
−5·0	23	73·4
−4·4	24	75·2
−3·9	25	77·0
−3·3	26	78·8
−2·8	27	80·6
−2·2	28	82·4
−1·7	29	84·2
−1·1	30	86·0
−0·6	31	87·8
0·0	32	89·6
0·6	33	91·4
1·1	34	93·2
1·7	35	95·0
2·2	36	96·8
2·8	37	98·6
3·3	38	100·4
3·9	39	102·2
4·4	40	104·0
5·0	41	105·8
5·6	42	107·6
6·1	43	109·4
6·7	44	111·2
7·2	45	113·0
7·8	46	114·8
8·3	47	116·6
8·9	48	118·4
9·4	49	120·2

50 to 100

°C	†	°F
10·0	50	122·0
10·6	51	123·8
11·1	52	125·6
11·7	53	127·4
12·2	54	129·2
12·8	55	131·0
13·3	56	132·8
13·9	57	134·6
14·4	58	136·4
15·0	59	138·2
15·6	60	140·0
16·1	61	141·8
16·7	62	143·6
17·2	63	145·4
17·8	64	147·2
18·3	65	149·0
18·9	66	150·8
19·4	67	152·6
20·0	68	154·4
20·6	69	156·2
21·1	70	158·0
21·7	71	159·8
22·2	72	161·6
22·8	73	163·4
23·3	74	165·2
23·9	75	167·0
24·4	76	168·8
25·0	77	170·6
25·6	78	172·4
26·1	79	174·2
26·7	80	176·0
27·2	81	177·8
27·8	82	179·6
28·3	83	181·4
28·9	84	183·2
29·4	85	185·0
30·0	86	186·8
30·6	87	188·6
31·1	88	190·4
31·7	89	192·2
32·2	90	194·0
32·8	91	195·8
33·3	92	197·6
33·9	93	199·4
34·4	94	201·2
35·0	95	203·0
35·6	96	204·8
36·1	97	206·6
36·7	98	208·4
37·2	99	210·2
37·8	100	212·0

100 to 490

°C	†	°F
38	100	212
43	110	230
49	120	248
54	130	266
60	140	284
66	150	302
71	160	320
77	170	338
82	180	356
88	190	374
93	200	392
99	210	410
100	212	413·6
104	220	428
110	230	446
116	240	464
121	250	482
127	260	500
132	270	518
138	280	536
143	290	554
149	300	572
154	310	590
160	320	608
166	330	626
171	340	644
177	350	662
182	360	680
188	370	698
193	380	716
199	390	734
204	400	752
210	410	770
216	420	788
221	430	806
227	440	824
232	450	842
238	460	860
243	470	878
249	480	896
254	490	914

500 to 1000

°C	†	°F
260	500	932
266	510	950
271	520	968
277	530	986
282	540	1004
288	550	1022
293	560	1040
299	570	1058
304	580	1076
310	590	1094
316	600	1112
321	610	1130
327	620	1148
332	630	1166
338	640	1184
343	650	1202
349	660	1220
354	670	1238
360	680	1256
366	690	1274
371	700	1292
377	710	1310
382	720	1328
388	730	1346
393	740	1364
399	750	1382
404	760	1400
410	770	1418
416	780	1436
421	790	1454
427	800	1472
432	810	1490
438	820	1508
443	830	1526
449	840	1544
454	850	1562
460	860	1580
466	870	1598
471	880	1616
477	890	1634
482	900	1652
488	910	1670
493	920	1688
499	930	1706
504	940	1724
510	950	1742
516	960	1760
521	970	1778
527	980	1796
532	990	1814
538	1000	1832

Locate temperature in middle column†. If in °C read the °F equivalent in the right hand column. If in °F read the °C equivalent in the left hand column.

To convert degrees Celsius to **degrees Fahrenheit**: $t_f = 1·8 t_c + 32$.

To convert degrees Fahrenheit to **degrees Celsius**: $t_c = \dfrac{t_f - 32}{1·8}$.

Temperature in degrees kelvin (K) equals temperature in degrees Celsius (°C) plus 273·15.

STEAM TABLES (SI Units)

The following tables have been extracted from *U.K. Steam Tables in SI Units 1970*, published by Edward Arnold, London.

Saturation Line

Temp., °C	Pressure, bar	Specific Volume, cm³/g		Specific Enthalpy, J/g	
		Water	Steam	Water	Steam
0	0·006 108	1·000 21	206 288	−0·041 6	2501
0·01	0·006 112	1·000 21	206 146	0·000 611	2501
10	0·012 271	1·000 4	106 422	41·99	2519
20	0·023 368	1·001 8	57 836	83·86	2538
30	0·042 418	1·004 4	32 929	125·66	2556
40	0·073 750	1·007 9	19 546	167·47	2574
50	0·123 35	1·012 1	12 045	209·3	2592
60	0·199 19	1·017 1	7 677·6	251·1	2609
70	0·311 61	1·022 8	5 045·3	293·0	2626
80	0·473 58	1·029 0	3 408·3	334·9	2643
90	0·701 09	1·035 9	2 360·9	376·9	2660
100	1·013 25	1·043 5	1 673·0	419·1	2676
110	1·432 7	1·051 5	1 210·1	461·3	2691
120	1·985 4	1·060 3	891·71	503·7	2706
130	2·701 1	1·069 7	668·32	546·3	2720
140	3·613 6	1·079 8	508·66	589·1	2734
150	4·759 7	1·090 6	392·57	632·2	2747
160	6·180 4	1·102 1	306·85	675·5	2758
170	7·920 2	1·114 4	242·62	719·1	2769
180	10·027	1·127 5	193·85	763·1	2778
190	12·553	1·141 5	156·35	807·5	2786
200	15·550	1·156 5	127·19	852·4	2793
210	19·080	1·172 6	104·265	897·7	2798
220	23·202	1·190 0	86·062	943·7	2802
230	27·979	1·208 7	71·472	990·3	2803
240	33·480	1·229 1	59·674	1037·6	2803
250	39·776	1·251 2	50·056	1085·8	2801
260	46·941	1·275 5	42·149	1135·0	2796
270	55·052	1·302 3	35·599	1185·2	2790
280	64·191	1·332 1	30·133	1236·8	2780
290	74·449	1·365 5	25·537	1290	2766
300	85·917	1·403 6	21·643	1345	2749
310	98·694	1·447 5	18·316	1402	2727
320	112·89	1·499 2	15·451	1462	2700
330	128·64	1·562	12·967	1526	2666
340	146·08	1·639	10·779	1596	2623
350	165·37	1·741	8·805	1672	2565
360	186·74	1·894	6·943	1762	2481
370	210·53	2·22	4·93	1892	2331
371	213·06	2·29	4·68	1913	2305
372	215·63	2·38	4·40	1937	2273
373	218·2	2·51	4·05	1969	2230
374	220·9	2·80	3·47	2032	2146
374·15 ±0·10	221·2	3·17	3·17	2095	2095

Notes: The specific internal energy is made exactly zero for the liquid phase at the triple point. The states here shown at 0 °C are metastable. At a pressure of exactly 1·013 25 bar the saturation temperature has the exact assigned value of 100 °C on the International Practical Scale of Temperature, 1948.

At a temperature of exactly 100 °C on the Thermodynamic Celsius Scale the saturation pressure is 1·013 25 bar, with a tolerance of 0·000 04 bar.

Near the critical point, the tolerances on the specific volume and on the specific enthalpy in the vapour phase are correlated with the corresponding tolerances in the liquid phase. The tolerances on the changes in specific volume and in specific enthalpy on evaporation tend to zero as the critical point is approached.

Water and Steam; Specific Volume, cm³/g

Pressure, bar	Temperature, °C									
	0	50	100	150	200	250	300	350	375	400
1	1·0002	1·0121	1696	1936	2173	2406	2639	2871	2987	3103
5	0·9999	1·0119	1·0433	1·0906	425·1	474·4	522·5	570·1	593·7	617·2
10	0·9997	1·0117	1·0431	1·0903	206·0	232·7	257·9	282·4	294·5	306·5
25	0·9989	1·0110	1·0423	1·0894	1·1556	87·0	98·9	109·7	114·9	120·0
50	0·9976	1·0099	1·0410	1·0878	1·1531	1·2495	45·34	51·93	54·90	57·76
75	0·9964	1·0088	1·0398	1·0862	1·1507	1·2452	26·71	32·44	34·75	36·91
100	0·9952	1·0077	1·0386	1·0846	1·1483	1·2409	1·397	22·44	24·53	26·40
125	0·9940	1·0066	1·0373	1·0830	1·1460	1·2367	1·387	16·14	18·25	20·01
150	0·9928	1·0055	1·0361	1·0813	1·1436	1·2327	1·378	11·49	13·91	15·65
175	0·9915	1·0044	1·0348	1·0798	1·1414	1·2288	1·369	1·716	10·57	12·46
200	0·9904	1·0033	1·0336	1·0782	1·1391	1·2251	1·360	1·665	7·68	9·95
225	0·9892	1·0023	1·0324	1·0766	1·1369	1·2215	1·352	1·630	2·49	7·86
250	0·9880	1·0012	1·0313	1·0751	1·1347	1·2179	1·345	1·600	1·98	6·00
275	0·9868	1·0002	1·0301	1·0736	1·1326	1·2144	1·338	1·576	1·865	4·19
300	0·9856	0·9992	1·0289	1·0721	1·1304	1·2111	1·331	1·555	1·797	2·82
350	0·9834	0·9972	1·0267	1·0692	1·1264	1·2046	1·319	1·519	1·705	2·111
400	0·9811	0·9951	1·0244	1·0664	1·1224	1·1984	1·308	1·489	1·644	1·912
450	0·9788	0·9932	1·0222	1·0636	1·1186	1·1925	1·297	1·464	1·599	1·804
500	0·9766	0·9912	1·0200	1·0609	1·1148	1·1868	1·288	1·443	1·564	1·731
550	0·9745	0·9892	1·0178	1·0582	1·1111	1·1813	1·278	1·424	1·533	1·677
600	0·9723	0·9873	1·057	1·0556	1·1075	1·1760	1·270	1·407	1·507	1·634
650	0·9703	0·9854	1·0137	1·0530	1·1040	1·1709	1·261	1·393	1·484	1·599
700	0·9682	0·9836	1·0116	1·0505	1·1006	1·1660	1·254	1·380	1·464	1·569
750	0·9662	0·9818	1·0096	1·0480	1·0973	1·1614	1·246	1·367	1·446	1·543
800	0·9642	0·9800	1·0076	1·0456	1·0941	1·1568	1·239	1·355	1·430	1·519
850	0·9622	0·9782	1·0057	1·0432	1·0910	1·1524	1·232	1·345	1·415	1·498
900	0·9603	0·9765	1·0038	1·0409	1·0879	1·1481	1·226	1·334	1·401	1·480
950	0·9584	0·9748	1·0019	1·0386	1·0848	1·1439	1·220	1·324	1·388	1·463
1000	0·9566	0·9731	1·0000	1·0363	1·0818	1·1398	1·214	1·314	1·376	1·447

Water and Steam; Specific Enthalpy, J/g

Pressure, bar	Temperature, °C									
	0	50	100	150	200	250	300	350	375	400
0	2502	2595	2689	2784	2880	2978	3077	3178	3229	3280
1	0·06	209·3	2676	2777	2876	2975	3074	3175	3227	3278
5	0·47	209·6	419·4	632·2	2857	2961	3064	3168	3220	3272
10	0·98	210·1	419·7	632·4	2830	2943	3051	3158	3211	3264
25	2·50	211·3	421·0	633·4	852·8	2881	3009	3126	3184	3240
50	5·05	213·5	422·8	634·9	853·8	1085·8	2925	3068	3134	3196
75	7·58	215·7	424·7	636·5	855·0	1085·9	2814	3003	3079	3149
100	10·1	217·9	426·6	638·1	856·1	1086·0	1343	2924	3017	3098
125	12·6	220·0	428·5	639·7	857·2	1086·1	1340	2826	2946	3041
150	15·1	222·1	430·4	641·3	858·3	1086·3	1338	2692	2861	2978
175	17·6	224·3	432·3	642·9	859·5	1086·5	1336	1663	2755	2905
200	20·1	226·5	434·2	644·5	860·6	1086·8	1334	1646	2605	2819
225	22·6	228·6	436·1	646·1	861·8	1087·3	1332	1633	1980	2715
250	25·1	230·7	438·0	647·7	863·0	1087·7	1331	1623	1850	2580
275	27·5	232·8	439·9	649·3	864·2	1088·2	1330	1615	1814	2383
300	30·0	235·0	441·8	650·9	865·4	1088·7	1329	1609	1791	2157
350	34·9	239·2	445·6	654·1	867·9	1090	1327	1598	1762	1992
400	39·7	243·5	449·4	657·4	870·4	1091	1325	1590	1743	1934
450	44·6	247·7	453·2	660·7	873·0	1092	1324	1582	1729	1901
500	49·3	252·0	457·0	664·0	875·6	1094	1324	1577	1717	1878
550	54·1	256·2	460·8	667·3	878·4	1096	1323	1572	1709	1860
600	58·8	260·4	464·6	670·6	881·1	1097	1323	1568	1702	1847
650	63·5	264·6	468·4	674·0	883·8	1099	1323	1565	1696	1836
700	68·1	268·8	472·1	677·3	886·6	1101	1323	1562	1691	1828
750	72·7	273·0	476·0	680·7	889·3	1103	1324	1560	1687	1820
800	77·3	277·1	479·8	684·0	892·2	1105	1324	1559	1684	1814
850	81·9	281·3	483·6	687·4	895·0	1107	1325	1557	1681	1808
900	86·5	285·4	487·3	690·8	898·0	1109	1326	1557	1678	1804
950	91·1	289·6	491·2	694·2	900·9	1111	1327	1556	1676	1799
1000	95·7	293·7	495·0	697·6	903·8	1114	1328	1555	1674	1796

425	450	475	500	550	Temperature, °C 600	650	700	750	800	Pressure, bar
3218	3334	3450	3565	3797	4028	4259	4490	4721	4952	1
640·6	664·1	687·5	710·8	757·4	803·9	850·4	896·9	943·2	989·6	5
318·4	330·3	342·2	354·0	377·5	401·0	424·4	447·7	471·1	494·3	10
125·0	130·0	135·0	139·9	149·6	159·2	168·8	178·3	187·7	197·2	25
60·53	63·24	65·89	68·50	73·61	78·62	83·6	88·4	93·3	98·1	50
38·96	40·93	42·83	44·69	48·28	51·76	55·16	58·52	61·82	65·09	75
28·12	29·73	31·26	32·76	35·61	38·32	40·96	43·55	46·09	48·58	100
21·56	22·98	24·31	25·59	27·99	30·26	32·44	34·56	36·64	38·68	125
17·14	18·45	19·65	20·80	22·91	24·88	26·77	28·59	30·35	32·09	150
13·93	15·19	16·31	17·36	19·28	21·04	22·71	24·31	25·86	27·38	175
11·47	12·71	13·79	14·78	16·55	18·16	19·67	21·11	22·50	23·85	200
9·51	10·76	11·81	12·76	14·42	15·92	17·31	18·62	19·88	21·10	225
7·89	9·17	10·22	11·14	12·72	14·12	15·42	16·63	17·79	18·91	250
6·50	7·85	8·90	9·79	11·32	12·65	13·86	15·00	16·08	17·11	275
5·298	6·736	7·799	8·682	10·16	11·43	12·58	13·64	14·65	15·62	300
3·430	4·956	6·054	6·928	8·340	9·516	10·56	11·52	12·42	13·27	350
2·546	3·686	4·758	5·620	6·980	8·086	9·051	9·93	10·75	11·52	400
2·191	2·916	3·814	4·628	5·934	6·982	7·885	8·70	9·45	10·16	450
2·010	2·492	3·170	3·884	5·114	6·108	6·960	7·72	8·42	9·07	500
1·896	2·245	2·750	3·342	4·464	5·404	6·209	6·93	7·58	8·19	550
1·816	2·085	2·474	2·950	3·950	4·831	5·592	6·27	6·89	7·46	600
1·756	1·976	2·283	2·672	3·543	4·360	5·080	5·72	6·31	6·85	650
1·706	1·892	2·144	2·466	3·221	3·971	4·648	5·26	5·81	6·32	700
1·665	1·828	2·040	2·310	2·965	3·648	4·283	4·86	5·39	5·87	750
1·631	1·775	1·958	2·189	2·760	3·380	3·972	4·52	5·02	5·48	800
1·602	1·731	1·892	2·092	2·594	3·155	3·706	4·22	4·70	5·14	850
1·576	1·693	1·837	2·014	2·458	2·966	3·478	3·97	4·42	4·84	900
1·552	1·660	1·790	1·948	2·344	2·806	3·282	3·74	4·17	4·57	950
1·530	1·630	1·750	1·892	2·248	2·670	3·111	3·54	3·95	4·34	1000

Note: The entry shown for 0 °C and 1 bar relates to a metastable liquid state. The stable state is here solid.

425	450	475	500	550	Temperature, °C 600	650	700	750	800	Pressure, bar
3332	3384	3436	3489	3597	3706	3817	3929	4043	4159	0
3330	3383	3435	3488	3596	3705	3816	3928	4043	4159	1
3325	3377	3430	3484	3592	3702	3813	3926	4040	4157	5
3317	3371	3425	3478	3587	3698	3810	3923	4038	4155	10
3295	3350	3406	3462	3574	3686	3799	3914	4030	4147	25
3257	3317	3375	3434	3550	3666	3782	3898	4016	4136	50
3216	3280	3342	3404	3526	3645	3764	3883	4003	4124	75
3172	3242	3309	3374	3501	3625	3747	3868	3990	4112	100
3125	3201	3273	3343	3476	3604	3729	3852	3976	4100	125
3073	3157	3235	3310	3450	3582	3711	3836	3962	4089	150
3017	3111	3196	3277	3423	3560	3692	3821	3949	4077	175
2955	3062	3155	3241	3396	3538	3673	3805	3935	4065	200
2885	3009	3112	3205	3368	3515	3654	3789	3922	4053	225
2807	2952	3066	3167	3339	3492	3635	3773	3908	4041	250
2718	2890	3018	3125	3308	3467	3615	3757	3894	4030	275
2614	2822	2967	3084	3278	3444	3596	3740	3880	4018	300
2375	2672	2858	2998	3216	3396	3557	3708	3853	3994	350
2203	2514	2741	2906	3153	3347	3518	3676	3826	3971	400
2115	2380	2624	2813	3088	3298	3478	3643	3798	3948	450
2064	2288	2522	2723	3023	3249	3439	3611	3771	3925	500
2030	2228	2439	2641	2960	3200	3400	3579	3744	3902	550
2005	2183	2378	2571	2900	3153	3362	3547	3718	3879	600
1986	2151	2330	2514	2844	3107	3324	3516	3692	3857	650
1971	2126	2294	2468	2793	3062	3288	3486	3666	3836	700
1958	2106	2265	2430	2748	3021	3253	3456	3641	3814	750
1948	2090	2241	2399	2709	2983	3219	3428	3617	3793	800
1938	2077	2222	2373	2674	2948	3187	3400	3593	3773	850
1932	2065	2206	2351	2644	2916	3157	3373	3570	3753	900
1925	2056	2193	2333	2618	2887	3129	3348	3548	3734	950
1920	2047	2181	2318	2595	2861	3103	3324	3527	3715	1000

Notes: The specific internal energy is made exactly zero for the liquid phase at the triple point.
The entry shown for 0 °C and 1 bar relates to a metastable liquid state. The stable state is here solid.

SI UNITS FOR VALVES

The quantities of greatest concern to the valve maker and valve user are those for:

length
mass
temperature
pressure
volume or capacity
stress

The SI units recommended by the BVMA for these quantities, for use in catalogues, sales literature, etc., together with the respective SI coherent units, are set out in the accompanying table.

Quantity	SI unit recommended by BVMA	SI coherent or primary unit
length	mm	m
mass	kg	kg
temperature	°C	K
pressure	bar (10^5 Pa)	Pa (1 Pa = 1 N/m²)
volume or capacity	l (litre)	m³
stress (UTS,YP)	MPa or N/mm²	Pa (1 Pa = 1 N/m²)

Note that only one of the BVMA recommended units, namely kilogram (kg), is in the coherent class of SI units.

Millimetre (mm) and litre (l) are decimal submultiples, and megapascal (MPa) is a decimal multiple, of coherent SI units and are recognized as such within the SI system. The bar is also a decimal multiple of a coherent SI unit but is not a selected multiple within SI. Nevertheless the bar is being retained because of its practical usefulness as a unit of pressure in some sectors of industry both in Britain and on the Continent.

The SI unit of temperature is the kelvin (K) and this unit is equally applicable to thermodynamic temperatures and to temperature differences. It is however generally recognized that the term degree Celsius (°C) will continue in use as the everyday customary unit of temperature. The units of Celsius and kelvin temperature interval are identical but the zero of the Celsius scale is the temperature of the ice point (273·15 K).

It is recognized that it will not always be practical to limit the usage of SI units to those recommended in the foregoing and other decimal multiples and submultiples will also be used. A list of SI prefixes used to form names and symbols of multiples of SI units is given in the table.

Factor by which the unit is multiplied	Prefix	
	Name	Symbol
10^{12}	tera	T
10^9	giga	G
10^6	mega	M
10^3	kilo	k
10^2	hecto	h
10	deca	da
10^{-1}	deci	d
10^{-2}	centi	c
10^{-3}	milli	m
10^{-6}	micro	μ
10^{-9}	nano	n
10^{-12}	pico	p
10^{-15}	femto	f
10^{-18}	atto	a

For full information on SI units and their application reference should be made to the relevant publications of the British Standards Institution.

INDEX

NOTES